Construction Management
New Directions

Second Edition

Denny McGeorge

Angela Palmer

with Kerry London

Blackwell
Science

Editorial offices:
Blackwell Science Ltd, 9600 Garsington Road, Oxford OX4 2DQ, UK
Tel: +44 (0) 1865 776868
Blackwell Publishing Inc., 350 Main Street, Malden, MA 02148-5020, USA
Tel: +1 781 388 8250
Blackwell Science Asia Pty, 550 Swanston Street, Carlton, Victoria 3053, Australia
Tel: +61 (0)3 8359 1011

First edition published 1997
Reprinted 2000
Second edition published 2002

3 2007

ISBN: 978-0-632-06042-9

Library of Congress Cataloging-in-Publication Data

McGeorge, W. D.
Construction management : new directions / Denny McGeorge, Angela Palmer ; with Kerry London.--2nd ed.
p. cm.
Includes bibliographical references and index.
ISBN 0-632-06042-5 (pbk.)
1. Construction industry--Management. I. Palmer, Angela. II. London, Kerry. III. Title.

HD9715.A2 M354 2002
624'.068--dc21

2002028020

A catalogue record for this title is available from the British Library

Set in 11/14 pt Bembo
by DP Photosetting, Aylesbury, Bucks
Printed and bound in Singapore
by C.O.S. Printers Pte Ltd

The publisher's policy is to use permanent paper from mills that operate a sustainable forestry policy, and which has been manufactured from pulp processed using acid-free and elementary chlorine-free practices. Furthermore, the publisher ensures that the text paper and cover board used have met acceptable environmental accreditation standards.

For further information, visit our subject website:
www.blackwellpublishing.com/construction

To
Jane, Paul, Graham, Liz and Tomo

Contents

Preface to second edition

In the preface to the first edition of *Construction Management: New Directions* we began with the rather bold statement that the book was meant to be topical. At the time of writing the first edition we realised that if the book was a success then inevitably there would be succeeding editions. The book has indeed proved to be topical enough to justify the production of a second edition, and we have found that writing the second edition has presented quite a different set of opportunities and challenges to the first edition. The primary reason for writing the first edition was the need to bring together, in a single volume, the concepts of *value management*; *total quality management*; *constructability*; *benchmarking*; *partnering* and *reengineering*. We still believe that the basic need exists to bring modern management concepts such as these to the fore if the construction industry is to achieve the cultural shift so long advocated by Latham[1], Gyles[2] and, more recently, Egan[3]. The adoption of concepts such as partnering is gathering momentum in, for example, Hong Kong, which is a recent adopter of partnering.

China, with its acceptance into the WTO (World Trade Organisation) and also with its recently successful bid for the Beijing Olympics in 2008, is well down the path of moving from a planned economy to a market economoy. Western approaches to construction management such as benchmarking and reengineering are generating considerable interest in China. Given that the number of workers currently employed in the Chinese construction industry is estimated at 34 million, there is considerable scope for adoption and adaptation of Western concepts to the Chinese version of a market economy.

One theme which underpinned the first edition was the need to recognise cultural differences when importing or transplanting management concepts. We identified distinct cultural differences between the United States, Europe and Australia in their approach to concepts such as value management, total quality management, partnering and reeingineering. By way of example, we described that when British companies tried to use the US system of value engineering they found initially that they could not make it work until the approach was modified to suit the UK professional environment. This is only one of many examples. It seems inevitable that the spread of Western management concepts into Asia will produce interesting sets of hybridised management concepts generated in response to differing cultural mores. Having recently observed the introduction of a 'tangible construction market'[4] throughout China, the flow of concepts may not only be from West to East but also in the reverse direction.

One deliberate omission from the first edition was the exclusion of information technology (IT) from our field of enquiry. At that point in time we felt that much was being written on the impact of IT (in all its various manifestations) on the construction industry. We also subscribed to the view that IT is an enabler of change rather than a change agent. We still hold to this view despite the fact that concepts such as supply chain management and reengineering rely heavily on IT for their application.

This second edition contains a new chapter on supply chain management. In the four years since writing the first edition, supply chain management has become a popular management tool and has found favour in many industries relying upon manufacturing production and distribution functions and has been used with a high degree of success. The fragmented nature of the construction industry and the perceived poor performance in productivity has prompted many to look to significantly better performing industries such as the automotive, retailing and information technology sectors to adopt better management practices. To some extent supply chain management can be traced back to the very origins of trading and commerce; however, its recent emergence as an explicit rather than an implicit management concept holds out exciting possibilities for the construction industry's complex and extended supply chain.

Another new area of coverage in this edition is the addition of a section on strategic alliancing. According to Doz and Hammel[5] one of the most dynamic features of modern corporate development has been the growth of alliances. The Economist Intelligence Unit[6] has expressed the view that alliances of all kinds will rank as one of the most significant management tools by the year 2010. In our view alliancing is a natural progression from partnering. In a successful strategic alliance the ethos of partnering is retained whilst at the same time contractual and commercial relationships are introduced which are binding on the stakeholders. In the light of these developments the chapter on partnering has been revisited and retitled 'Partnering and alliancing'.

In the first edition the concept for which we had least empathy was reengineering. Perhaps this was because of the extravagant claims made by its protagonists such as 'You can choose to reengineer or you can choose to go out of business'[7] or Hammer's often quoted saying 'Don't automate, obliterate'[8]. Certainly this is a concept where there is a distinct cultural divide between the US and Europe. We believe, however, that there is merit in its retention in this second edition. In a text which focuses more on management concepts than on management techniques, the 'big bang' discontinuous improvement approach of reengineering certainly has a place, even if it is at the far end of the spectrum. In the first edition we explored how systems theory underpins many modern management concepts. The somewhat conciliatory stakeholder approach of systems theory may seem to be out of place with reengineering methods which include appointing a reengineering 'czar' (champion); however, beneath the hype there is clear evidence of the systems approach at work.

There is also clear evidence of the systems approach in lean construction[9]. Lean construction is also influenced by total quality management (TQM) and partnering and supply chains. It is a relatively new concept, the Lean Construction Institute (LCI) having been founded in August 1997, thus postdating the first edition of this book. The claims made in favour of lean construction are substantial, if less boastful, than those for reengineering. Lean construction is closely linked to supply chain management and for this reason is discussed in the context of the supply chain management chapter.

At the beginning of this preface we stated that writing a second edition inevitably brings with it a new set of opportunities and challenges. We have outlined the opportunities which the second edition has given us to review current management concepts (many of the new inclusions were prompted by the helpful comments we received on the first edition). The ever-present challenge has been for us to remain as detached and objective as possible. As in the first edition, the purpose of this book is not to champion any particular management concept in preference to another. Indeed, many of the emerging concepts seem well able to champion their own cause. The problem for the decision taker still remains in being able to relate to the many construction management concepts which, at times, appear to be in competition with one another for the decision taker's attention. The purpose of this book is to provide a framework in which concepts are viewed as synergistic rather than mutually exclusive.

References

1. Latham, M. (1994) *Constructing the Team*. HMSO, London.
2. Gyles, R.V. (1992) *Royal Commission into Productivity in the Building Industry in New South Wales*. Government of New South Wales, Sydney.
3. Egan, J. (1998) *Rethinking Construction*. DETR, London.
4. McGeorge, D. & Zou, P. (2001) A perspective on construction management and economics issues in Western and Chinese Construction Industries. *2001 International Conference on Project Cost Management* (eds Liu, A.M.M., Fellows, R. & Drew, D.). Ministry of Construction, China Engineering Cost Association and the Hong Kong Institute of Surveyors, Beijing, May 2001.
5. Doz, Y.L. & Hammel, G. (1998) *Alliance Advantage; The Art of Creating Value through Partnering*: Harvard Business School Press, New York.
6. Economist Intelligence Unit (1997) *Vision 2010; Designing Tomorrow's Organisation*. Economist Intelligence Unit, London, New York, Hong Kong.
7. Morris, D. & Brandon, J. (1993) *Re-engineering your Business*. McGraw-Hill, New York.
8. Hammer, M. & Champy, J. (1993) *Reengineering the Corporation: A Manifesto for Business Revolution*. Nicholas Brealey, London.
9. Howell, G. (1999) What is lean construction? *Proceedings IGLC-7*. University of California, Berkeley, CA, 26–28 July, 1–10.

Acknowledgments

We would like to thank Takayaki Minato of Tokai University Japan and John Kelly at Glasgow Caledonian University, Scotland for their value management case studies outlined in Chapter 2; Professor Vernon Ireland for allowing us to use the T40 case study material, Dr Selwyn Tucker and his colleagues at CSIRO, Victoria, Australia for allowing us access to their extensive writings on reengineering; Professor Tony Sidwell and his colleagues at the Construction Industry Institute, Australia for their input both on constructability and partnering; Glen Peters for allowing use of his benchmarking survey; and Chen Swee Eng, Rod Gameson and colleagues at the University of Newcastle for their input on constructability, decision support systems and alliancing.

In particular, we would like to thank Kerry London for the invaluable contribution of her knowledge of supply chain management.

The nature of this book has meant that we have drawn on a large number of other sources and we acknowledge our indebtedness to many commentators in construction management and management science who are too numerous to mention by name, but without whom this book could not have been written.

Chapter 1

A cultural shift in the construction industry

For many years the construction industry has been criticised for its perceived inability to innovate and its slow adoption of new technology and modern management methods. The industry has long been exhorted to change its ways and, in the UK in particular, there has been a seemingly endless procession of reports and enquiries ranging from the Simon Report[1], the Emmerson Report[2] and the Banwell Report[3], the Latham Report[4], and, more recently, the Egan Report[5]. In addition, reports such as *A Fresh Look at the UK and US Building Industries*[6], *Controlling the Upwards Spiral: Construction Performance and Cost in the UK and Mainland Europe*[7], *Building Britain 2001*[8], *Strategies for the European Construction Sector: A Programme for Change*[9], have increased pressure for change in the UK construction industry. In the southern hemisphere pressure has been exerted through the Gyles *Royal Commission into Productivity in the Building Industry in New South Wales*[10] and the Construction Industry Development Agency's *Reform Strategy*[11], both of which preceded Latham.

Powell[12], in his study of the economic history of the British building industry 1815 to 1979, comments, not unkindly, that the period of 1940 to 1973 was when the 'work horse learned to canter'. Perhaps, to continue Powell's analogy, the period of the 1990s is where, as a result of vigorous prompting by observers such as Egan, Latham and Gyles, the workhorse has moved from a canter to a gallop.

Many of the concepts dealt with in this book are directly advocated by Egan, Latham and Gyles. For example, in respect to partnering, Latham[4] comments that:

'Specific advice should be given to public authorities so that they can experiment with partnering arrangements where appropriate long-term relationships can be built up. But the partner must initially be sought through a competitive tendering process, and for a specific period of time. Any partnering arrangement should include mutually agreed and measurable targets for productivity improvements.'

In Australia, Gyles went further than this and initiated a pilot partnering project through the aegis of a royal commission.

It would be difficult, even for the sharpest critic of the construction industry, to dispute that the management concepts covered in this book are not in evidence in the construction industries in both the northern and southern hemispheres. It would also be fair to say however that none of these concepts is, as yet, commonplace in the industry. The concepts of value management, total quality management, constructability, benchmarking, partnering and reengineering would have been unheard of to either a busy construction manager or most construction management undergraduates in the decade of the 1980s and will still be unfamiliar to many, even at the beginning of the twenty-first century.

The dominant message from both Latham's final report and Gyles's royal commission report is the key role of the client in activating a cultural shift in the industry through the adoption of modern management concepts. This is summarised by Latham who states that

'implementation begins with clients. Clients are at the core of the process and their needs must be met by industry'.

Latham then goes on to recommend that

'Government should commit itself to being a best practice client. It should provide its staff with the training necessary to achieve this and establish benchmarking arrangements to provide *pressure* for continuing improvements in performance.' (our italics).

This has been expressed more succinctly, if more crudely, as 'the client having the power of the cheque book'.

In his report, Latham makes mention of the Australian approach to cultural change through the Construction Industry Development Agency. CIDA was an Australian Federal Government initiative, set up under the Construction Industry Reform and Development Act of 1992. CIDA's remit was to bring about a 'real and measured change'. CIDA had a fixed life which expired in June 1995. A legacy of CIDA has been, among other things, a Code of Practice of prequalification criteria. Although CIDA no longer exists, Australian State Government Authorities such as the New South Wales Department of Public Works and Services (DPWS) continue to actively promote best practice through the use of contractor prequalification schemes using prequalification criteria.

The DPWS contractor accreditation scheme[13] lists the following construction industry best practice initiatives:

(1) Commitment to client satisfaction;
(2) Quality management;
(3) Occupational health and safety and rehabilitation management;
(4) Co-operative contracting;
(5) Workplace reform;
(6) Management of environmental issues;
(7) Partnering;
(8) Benchmarking;
(9) Another area of best practice nominated by the contractor and accepted by DPWS.

For contracts valued at over Aus. $20 million contractors must have a demonstrated record of commitment to, or a corporate programme for, the early implementation of criteria 1 to 6 and at least one of the reform initiatives 7 to 9.

Under the scheme, contractors who achieve best practice accreditation will be offered significantly more tendering opportunities than those contractors who are not accredited. It is anticipated that ultimately, only those contractors accredited under the scheme will be eligible for selection to tender for contracts valued at over Aus. $500,000. (A good example of the power of the cheque book.)

Of the seven concepts covered in this book, three are explicitly listed by DPWS. These are quality management, partnering and benchmarking. Value management is not listed as an optional criterion, because it is a DPWS mandatory requirement on all major projects. Given that under section 9 contractors are allowed to nominate another area of best practice (for example, constructability) then most of the management concepts contained in this book are essential criteria for contractors wishing to prequalify for tendering for government contracts in New South Wales.

The DPWS example is given, not by way of promoting this book, but by way of illustrating its central theme: that clients are taking on board the directives of Egan, Latham and Gyles and their predecessors and are now committed to continuous improvement through the introduction of modern management concepts. As previously stated, in the decade of the 1980s the practising construction manager or the construction management undergraduate would not have heard of concepts such as value management, total quality management, buildability/constructability, benchmarking, partnering and reengineering. In the twenty-first century there is a greater awareness of these concepts, and clients, particularly government agencies, are pushing hard for their adoption. However there is also a good deal of ambiguity about the nature and application of the concepts[14].

There are probably several reasons why many of the concepts in this book are not clearly understood and hence have yet to be adopted by the industry at large. One reason may simply lie in the fact that hitherto a balanced description of the concepts has not been presented in total, and we have tried to redress this. Another and more complex reason may lie in the fact that many if not all of the concepts under consideration are philosophically grounded, if not in systems theory, then at least in a holistic approach. We would contend that this common parentage has given rise to difficulties in terms of identifying the concepts as individual branches of the same family tree. This lack of differentiation between current concepts is typified in comments such as 'constructability is not just value engineering or value management'[15] or 'is reengineering replacing total quality?'[16] or more confusingly, 'as partnering is to the project, total quality management is to the construction company'[17].

For many years critics of the construction industry have dwelt on the perceived problems of fragmentation and compartmentalisation. Many of the ills which have beset the industry have been blamed on the inability of the industry to see the big picture. Many of the advocates of the techniques covered in this book claim that 'their' concept rectifies this. For example Hellard[18] advocates that

> 'Partnering will certainly be the key to the *holistic* approach which must first be brought to the organisation and then incorporated into the team performance with other sub-contractors and the main contractor.'

We find no fault with these sentiments; however most of the other current construction management concepts would also subscribe to similar sentiments. The result of this convergence of ideals is that many construction management concepts appear to be in competition with one another for the attention of the same decision-makers.

This can be illustrated by reference to the 'cost influence curve' (based on the Pareto Principle) which has been used extensively in construction management literature to demonstrate that the earlier an individual or group is involved in the decision-making process, the greater the potential for impact on the project outcome. Conversely the ability to influence the project outcome diminishes exponentially over time. The problem lies in the conflict which can arise when a large number of concepts compete with one another for the prime position at the origin of the x, y axis (Figure 1.1).

The probability is that the relatively slow and patchy uptake of modern construction management concepts is due not so much to a lack of diligence or a reluctance on the part of industry practitioners to adopt new ideas, but to the fact that these concepts need first to be understood and studied in total. Second, although government agencies are encouraging and, in some cases, attempting to enforce their adoption of the concepts, no guidance is being given on how the concepts should be applied concurrently and in combination. What is needed is a *Weltanschauung* or worldview based on a solid knowledge of the individual concepts. This is the purpose of this book.

The book's contents

Value management

Value management was developed in the United States manufacturing industry during World War II. Its aim was to improve the value of goods by concentrating on the functions that products perform. It was so successful in manufacturing that the United States Department of Defense began using it in the construction industry and it was around this time that an interest in value management was shown by the British construction industry. Chapter 2 traces the historical development of value management; the use of function analysis; organisation of value management studies; the evaluation of

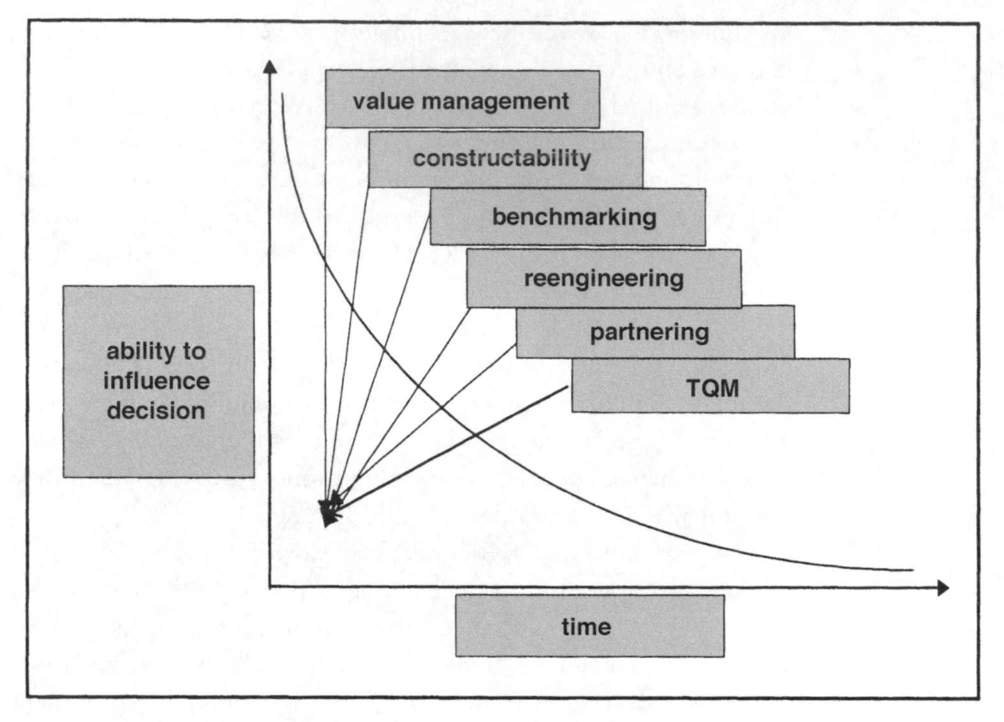

Fig. 1.1 Conflicting demands at origin of the cost influence curve.

value management proposals; the American system of value management; the British system of value management; and the Japanese system of value management. The chapter ends by analysing why these three systems are different and examines some of the major cultural influences on value management development.

Constructability

Constructability is the only concept in this book which is the exclusive domain of the construction industry. Constructability is concerned with how decisions taken during the procurement process facilitate the ease of construction and quality of the completed project. From its inception in the early 1980s constructability has moved from its original narrow focus to incorporate decision support theory and decision support systems. The following aspects of constructability are covered: the origins; scope and goals; implementation; constructability in practice; the building – in use; good and bad constructability – indicators of success; and quantifying the benefits of constructability. Chapter 3 concludes by distinguishing between constructability and good multi-disciplinary team-working.

Benchmarking

Benchmarking is a concept aiming at improving the competitiveness of organisations through the examination and refinement of their business processes. The concept has its origins in the Xerox Corporation, who stripped down copiers manufactured by competitors and compared them to their own. They later extended this comparison to include the business processes of their competitors. Chapter 4 looks at types of benchmarking; the process of benchmarking; the benchmarking team and the benchmarking code of conduct. The chapter concludes by illustrating a simple case study of benchmarking the customer focus of a national housebuilder against a national car manufacturer.

Reengineering

Reengineering is being hailed as a management revolution, which could have repercussions on the scale of the industrial revolution which followed Adam Smith's *Wealth of Nations*. The proponents of business process reengineering are claiming quite dramatic results following its introduction. The full impact of reengineering is yet to be felt in the construction industry, although interest is gaining ground. The following aspects of reengineering are covered: origins; reengineering in a construction industry context; goals; methodology; implementation; time and cost saving; pitfalls; IT and reengineering; a European perspective; and the T40 project. Chapter 5 contains a detailed case study, known as the T40 project, of the initiation, planning and implementation of a process reengineering in the Australian construction industry. The objective of the project was the reduction of construction process time by 40%.

Total quality management

Total quality management or TQM is a concept aimed at improvement of the organisation through increased customer focus, integration of the organisation's processes and a philosophy of continuous improvement. Chapter 6 examines definitions of TQM; historical development; the need for a cultural change in the construction industry; customer focus; integration; continuous improvement; quality costs and quality standards. Finally, Chapter 6 briefly examines the array of quality methods that are currently available.

Supply chain management

Chapter 7 describes how the application of the supply chain concept to construction is now being advocated as a management strategy to improve the performance of the industry. It has found favour particularly with the research community and various government policy making units worldwide. The application of the concept has had mixed usage among construction industry practitioners. However, this appears to be changing, with ad hoc examples of industry participants such as clients, consultants, contractors, specialist subcontractors and major suppliers engaging in various forms of supply chain management behaviour, particularly in developing more long-term strategic relationships. There have been some notable applications by construction firms, however very little is known about the application of supply chain management in a widespread manner across the industry. Supply chain management is a complex and difficult subject; it is however likely to be one of the dominant fundamental concepts of doing business in the twenty-first century.

Partnering and alliancing

The concept of formal partnering is of relatively recent origin, dating back to the mid-1980s. The concept was developed in the United States and has spread to other countries including Australia and New Zealand in the southern hemisphere and also to the UK. Parties adopting partnering resolve to move away from the traditional adversarial relationships to a 'win-win' situation. Partnering can either be undertaken at the level of a single project and be of relatively short duration or can be of a semi-permanent nature at a strategic level. Chapter 8 traces the origins of partnering; partnering in a construction industry context; the goals of partnering; categories of partnering, project and strategic; the participants; commitment; the partnering process; how to conduct partnering workshops; partnering charters; the pitfalls of partnering; limits to partnering; legal and contractual implications of partnering; and dispute resolution. A substantial part of Chapter 8 is devoted to the exploration of 'alliancing' as a natural progression from partnering. Alliancing has been hailed as the one of the most dynamic features of modern corporate development. However its uptake by the building industry (as opposed to the oil and civil engineering industries) has been slight. It could be argued that alliancing combines the cultural features of partnering with the cutting edge of economic rationalism. Time will tell whether or not alliancing replaces or becomes a natural adjunct to partnering.

Linking the concepts

Chapter 9 explores the relationships between the various concepts. The case is made that the current raft of construction management concepts owes its parentage to systems theory and a systems approach. Rather surprisingly the use of the systems approach has not had the effect of bringing the concepts together, but, in fact, has had the opposite effect. A conceptual model is proposed which illustrates the relationships of the concepts, one to the other, based on the use of 'gentle guidelines' from soft systems theory.

The chapters are arranged in a roughly chronological order, although, as readers will soon discover, the precise dating of the emergence of a concept is sometimes difficult, and in any event is usually of no particular significance. Each chapter can be read as a stand-alone topic, but like all authors we would like to think that the book will be read from cover to cover. Certainly the arguments developed in Chapter 9 will only consolidate after reading Chapters 2 to 8.

References

1. Central Council for Works and Buildings (1944) *The Placing and Management of Building Contracts*. HMSO, London.
2. Emmerson, Sir H. (1962) *Study of Problems before the Construction Industries*. HMSO, London.
3. Committee on the placing and management of building contracts. (1964) *Report of the Committee on the Placing and Management of Building Contracts*. HMSO, London.

4. Latham, M. (1994) *Constructing the team*. HMSO, London.
5. Egan, J. (1998) *Rethinking construction*. DETR, London.
6. Flanagan, R., Norman, G., Ireland, V. & Ormerod, R. (1986) *A Fresh Look at the UK and US Building Industries*. Building Employers Confederation, London.
7. The Business Round Table. (1994) *Controlling the Upwards Spiral: Construction Performance and Cost in the UK and Mainland Europe*. The Business Round Table, London.
8. University of Reading. (1988) *Building Britain 2001*. Centre for strategic studies in construction, London.
9. European Commission (1994) *Strategies for the European Construction Sector: A Programme for Change*. KHL Publishing, Wadsworth, England.
10. Gyles, R.V. (1992) *Royal Commission into Productivity in the Building Industry in New South Wales*. Government of New South Wales, Sydney.
11. Construction Industry Development Agency (1992) *Construction Industry: In-principle Reform and Development Agreement, Reform Strategy*. AGPS, Canberra.
12. Powell, C.G. (1980) *An Economic History of the British Building Industry*. Methuen, Cambridge, England.
13. New South Wales Department of Public Works and Services. (1995) *Contractor Accreditation Scheme to Encourage Reform and Best Practice in the Construction Industry*. Government of New South Wales, Sydney.
14. Fong, P.S.W. (1996) VE in construction: a survey of clients' attitudes in Hong Kong. *Proceedings of the Society of American Value Engineers International Conference*, Vol. 31.0.
15. Sidwell, A.C. & Francis, V.E. (1995) The Application of Constructability Principles in the Australian Construction Industry. *Proceedings of the CIB W65 Conference*, Glasgow.
16. Kelada, J.N. (1994) Is re-engineering replacing total quality? *Quality in Progress*, Dec., 79–83.
17. Deffenbaugh, R.L. (1996) Why one contractor calls partnering project quality planning. In: *Partnering in Design and Construction* (ed. K. A. Godfrey Jr.) McGraw-Hill, New York 231–44.
18. Hellard, R.B. (1995) *Project Partnering*. Thomas Telford, London.

Chapter 2

Value management

Introduction

Most people would agree that there is a difference between cost and value and that, of the two, value is much more difficult to define. The value of a child's battery-operated radio in a modern home equipped with satellite TV and sophisticated hi-fi may be very small. When a snowstorm cuts the electricity supply and the small battery-operated radio is the only means of hearing news, its value is drastically increased.

Even *Chambers English Dictionary*[1] has several definitions of value, ranging from 'that which renders anything useful or estimable' to, more simply, 'price'. It is because of these different concepts embodied within the word value that value management is sometimes difficult to understand and it is possibly for this reason that it has been confused with cost-saving, buildability and cost-planning[2]. Such confusion has, not surprisingly, led to a fragmentation in the use of value management and in today's construction industry the technique is at the crossroads of its development[3]. Following an initial flurry of enthusiasm in the late 1980s, use and interest in the technique diminished, largely because there was a lack of appreciation of the different concepts that can exist within the term value. This can be clearly seen in the historical development of value management from its invention in the 1940s to its use in today's construction industry.

Historical development

Value engineering was the term first used by Miles[4] to describe a technique that he developed at the General Electric Company during World War II. The technique began as a search for alternative product components to cope with shortages which had developed as a result of the war. Because of the war, however, these alternative components were often equally unavailable. This led to a search not for alternative components but for a means of fulfilling the *function* of the component by an alternative method. It was later discovered that this process of function analysis produced cheaper overall products without reducing quality, and after the war the system was maintained as a means both of removing unnecessary cost from products and of improving design.

The central feature of Miles' work was the definition of all functions that the customer required of the product. These functions were defined only in terms of one verb and one noun. Miles believed that if such a definition could not be achieved, the real function of the item was not understood. The defined functions were then evaluated in terms of the lowest possible cost to achieve them and this evaluation was then used as a means of finding alternatives that also fulfilled the functions. Miles illustrated his work with an example of an electric motor screen which needed to perform the functions shown in Table 2.1:

Table 2.1 Function definition.

	Verb	Noun
1	Exclude	substance
2	Allow	ventilation
3	Facilitate	maintenance
4	Please	customer

These functions were then evaluated by selecting the cheapest possible means of achieving them. These were as follows:

- The function of 'exclude substance' was evaluated on the basis of the cost of a sheet of metal to shield the motor.
- The function of 'allow ventilation' was based on the additional cost of putting holes in the sheet metal.

- 'Facilitate maintenance' was evaluated by adding the cost of a spring clip to allow the sheet metal to be removed.
- 'Please customer' was based on the cost of painting the metal.

The costs of these are shown in Table 2.2.

Table 2.2 Allocating cost to function.

	Verb	Noun	Cheapest means of achieving function	Lowest cost to achieve function
1	Exclude	substance	Sheet metal	$0.15
2	Allow	ventilation	Holes in metal	$0.15
3	Facilitate	maintenance	Spring clip	$0.10
4	Please	customer	Paint metal	$0.10
Total lowest cost to achieve all functions				$0.50

This total lowest cost can therefore be viewed as the true value of the screen, since it is only this cost that needs to be incurred in order to meet all the functions that the customer requires. In going through this process Miles costed the functions of the screen based on the lowest possible cost of achieving them; in this case $0.50. This lowest cost was then compared to the actual cost of the existing screen: $4.75. This clearly highlighted that much of the cost incurred in producing the screen actually achieved no function.

A by-product of defining and evaluating function is the ease with which it allows further means of achieving the function to be generated. This can be illustrated with a construction example. In a hotel development the architect had included a child's paddling pool next to the main hotel pool. The function of this pool was not, as might first be thought, to provide a leisure activity for children. It was in fact a safety measure to keep them out of the main pool. Once this function of the child's pool was defined as 'keep (children) safe' it was much easier to generate alternatives that satisfied this function. A play area, a small playground or even a crèche would all meet the requirements. In the end the design team settled for a water spray which fulfilled the function of safety for much less cost than the original paddling pool. This function definition, function evaluation and generation of alternatives are collectively called function analysis and it is this basic technique which forms the basis of value engineering.

Clearly in complex organisations a system would be required to carry out function analysis: it could not be done on an ad hoc basis. When the function analysis would be carried out, who would do it, and how it would be organised would need to be considered. For this reason systems of value engineering, as it now came to be called, grew up around the technique of function analysis[5]. Figure 2.1[6] shows a typical system that developed. Central to the system was the concept of function analysis. The studies were organised into a series of steps known as the job plan and were carried out by a value engineering team at some time during the product's life cycle.

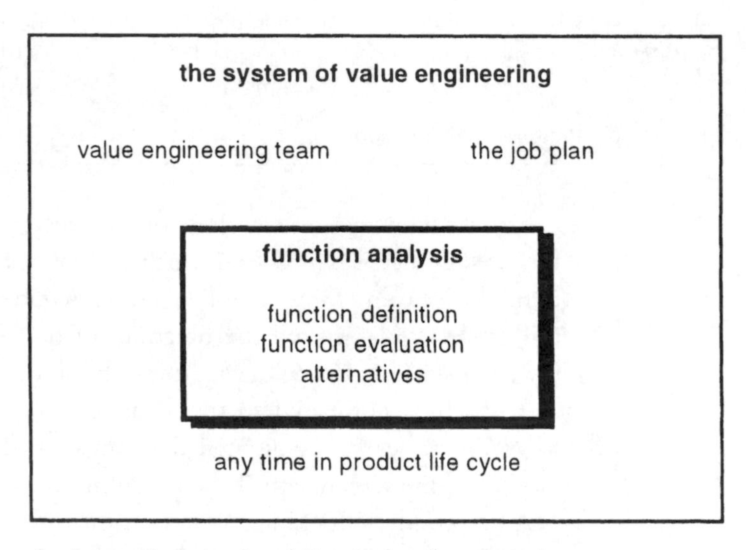

Fig. 2.1 The system of value engineering in relation to function analysis.

Up until the 1970s value engineering was only being used in the manufacturing sector but at around this time it started to be used in the United States for construction projects. Use was mainly by government agencies[3] who for various reasons adapted the manufacturing value engineering systems and tailored them for their own use.

As shown in figure 2.2 the 40-hour workshop became a feature of value engineering when it began to be used in the construction industry. In addition value engineering was carried out at the 35% design stage using a team external to the project; that is, a team not forming part of the design team. It was largely this system of value engineering, imported from the United States, that was first used in the UK construction industry.

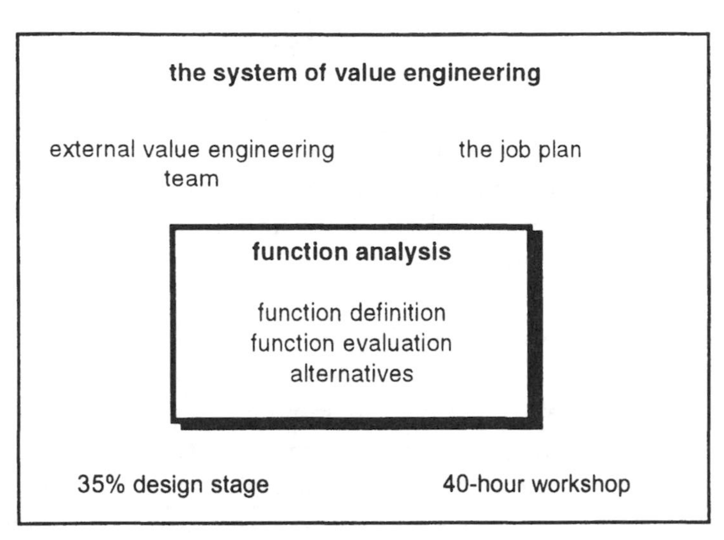

Fig. 2.2 Systems of value engineering in construction.

The problem with this American system, which is still widely used in the US, is that it deviates from the original work of Miles. It does not, other than in name, include function analysis. Value engineering (VE) in the US is basically a design audit. It consists of a 40-hour workshop structured loosely around a job plan. It is carried out at 35% design by an external team. It involves the selection of high cost areas and the generation of alternatives to them. The selection of high cost areas is a fairly loose procedure. It is based on the comparison of elemental costs with the cost of cheaper alternatives, along with a more general analysis of cost centres of the project. This nebulous approach results in a fairly broad VE output encompassing design changes and cost cuts from all disciplines. This output however cannot be attributed to function analysis. The actual workshop itself as an autonomous unit is a critical contributory factor in the success of value engineering studies. Within the workshop, the degree of success of the study relates largely to the personalities involved (particularly that of the leader), the timing of the study, the interaction of the VE team, the input of the design team and the role of the client. The technique of function analysis bears little or no relationship to the output of the study.

When British companies tried to use the US system of value engineering they found that they could not make it work. There were many reasons for this but they were chiefly related to the original objectives of the value engineering studies. The US system of value engineering was born out of a need for greater accountability on government projects. (Almost all value engineering activity in the US is government work.) The situation in the UK was very different. The quantity surveying system provided all the accountability that was needed. Value engineering was required to provide a platform for the examination of value as opposed to cost. In this light it was hardly surprising that the US system of value engineering broke down in the UK.

Once this breakdown had occurred the UK construction industry was faced with two options. It could abandon value engineering altogether or it could go back to the original work of Miles and build its own system that satisfied its own objectives. It appears that the latter route has been chosen and that new systems are developing under the title of value management. For this reason only this terminology will be used in the text from this point onwards. What this choice means is that the UK is essentially starting with a clean sheet in building new systems of value management. However, the US experience has shown that in addition to function analysis there are other components of value management systems, such as the team or the timing of the study, that influence the success or otherwise of value management. To build an effective value management system these components need to be investigated and the next section of this chapter examines these components in detail.

Before we move on, a final point regarding the American system needs to be clarified. The authors have defined value management as incorporating function analysis. This is our definition. As *Chambers Dictionary* illustrated, the concept of value may cover many things. The US system, despite neglecting function analysis, is useful and fulfils the objectives for which it is required.

Function analysis

On a recent visit to my new local surgery the nurse complained about the construction of the facility. In the room in which we were sitting, the door was directly opposite the work cabinets and this meant that the patient's trolley bed could not be pushed into or out of the room.

It was necessary for a patient to be wheeled in on a wheelchair and then lifted on to the trolley. It would, she complained, have been much easier if the door had been placed on the other side, allowing the trolley to be moved in and out freely without obstruction from the work cabinets.

Clearly in designing the room the designer did not consider what the room would be used for. He (or she) did not know that the room would be used to examine patients who were incapable of walking into the room and climbing on to the bed themselves; that is, he did not establish the function of the room correctly. The most likely reason that he did not do this was because he failed to involve in the design process the people who would use the room. As a result the value of the room was diminished, in that it did not properly perform the function for which it was required.

This failure to provide buildings or parts of buildings which properly perform their functions is a common problem in the construction industry and it is this that forms the basis of function analysis. In the context of function analysis it is assumed there is a close relationship between the provision of function and the achievement of value. Where all functions are accomplished at the lowest achievable cost there is good value. Where no function is achieved, or where function is achieved at too great a cost, there is little or no value.

The problem that arises at this point is that even if the designer can define the functions that a particular item is required to perform, how can he or she know that the functions are in fact being provided at the lowest achievable cost? Function evaluation takes place after function definition, and assesses the defined functions based on the lowest possible cost to achieve them. As an example think about a window, which in a particular location may have the sole function of allowing light to enter (admit light). The cheapest possible means by which this can be achieved is by forming an opening in the wall. The cost of doing this is therefore the *value* of the function. Naturally this example must not be taken literally. It is highly unlikely that the sole function of a window is to admit light. The example however illustrates that anything spent above the cost of achieving the function is unnecessary and reduces the value of the item. Practical illustrations of this example can be seen in many hot climates where the function of admitting light is in fact achieved solely by forming a hole in the wall.

A further benefit of function evaluation is that it is creative. It acts as an aid to finding other solutions that also satisfy function. Looking for the lowest possible cost to achieve function tends to lead to the question 'if a hole in the wall will satisfy the function what else will?' A hole in the roof, a see-through wall, or an electric lamp will all equally satisfy the function of admitting light.

As summarised in figure 2.3, the three items of function definition, function evaluation and creative alternatives are collectively called function analysis, and it is this that forms the basic core of value management. Without this technique, any exercise that is carried out on a project, useful though it may be, cannot be described as value management. For this reason each stage of the technique is examined in greater detail.

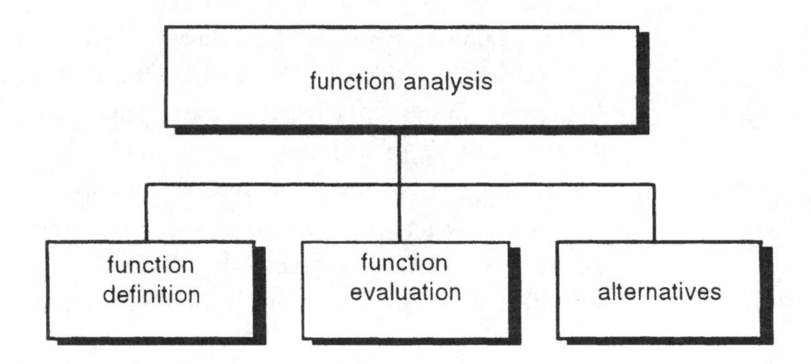

Fig. 2.3 The stages of function analysis.

Function definition

Function is defined through the use of a verb and a noun because this clarifies the function. Generally if the function of an item cannot be defined in terms of a verb and a noun, its function is not fully understood.

The problem that arises with function definition in construction is one of level. In the example the function of an element of the building, the window, was defined but it is also possible to break the window into a series of components which could also be defined, e.g. the head, sill or locks. Equally the function of the spaces provided by the project could be defined. In a school, for example, this might include the functions of classrooms and staffrooms. Alternatively it is possible to step back still further and define the function of the school as a whole. Defining the functions at these four different levels would produce four different results. Which one of these levels should therefore be chosen?

Defining the function of the project as a whole

Many projects are not what they seem. As an example think about the functions of a bridge, the most obvious function of which is to 'convey traffic'. However this may not necessarily be the case and the real function may be to take industry to another part of the city, or to provide work for the construction industry or to reduce traffic congestion. The correct definition of the project function at this level can be useful because if the correct function is, for example, to move industry to another part of a city, then it is clear that this can be achieved other than through the construction of a bridge. The offering of grants or subsidies is an obvious alternative.

Defining the functions of the spaces within the project

Once it has been decided that a particular project is required, the functions of the spaces within it can be defined. In a school, for example, we can ask what is the function of classrooms, playgrounds, staffrooms or sports facilities. The function of the playground might be to give the children an opportunity to let off steam, to allow them some fresh air, or to give the teachers a well-earned break. Depending on the function, the design solution, and any viable alternative to it, will vary.

Defining the function of the elements

In addition to project and space function, the function of project elements can also be defined. Think once again about a window. If you were asked you would probably define its functions as 'provide ventilation' or 'admit light'. But now think about the function of a window in a prison cell. In this case the function is not to admit ventilation as the window will rarely, if ever, open. Nor will the function be to admit light, as often the amount of light provided is insufficient even for day use. The real function of the window is to 'humanise environment'. The definition of elemental functions can, therefore, also provide a clearer insight into design and design alternatives. However it is not possible to define the functions of elements generically, without reference to the building itself. It can never be assumed that the function of internal walls is to 'divide space'. The function of internal walls in a prison cell is entirely different from the function of internal walls in a toilet block. Neither have the function of dividing space.

Defining the function of components

It is possible to divide elements into components and once again this can provide insight into design and design alternatives. In breaking a window into components of sills, heads, and jambs, these parts can be examined to see how well they meet the required functions.

Which level of function should be selected?

Given that there are four possible levels of function definition, which one should be used for a construction project? The answer to this depends on the client, the level of design completed and the project generally. Where the client is adamant that the project concept is the one required, there is little purpose defining its function and a function definition on space may be the most appropriate. Likewise if the spaces provided are fixed then an elemental function definition may be the best way of improving value. Generally in construction function definition is not carried out on components as this is viewed as the remit of the manufacturer.

It is of course possible to do all three or any two of these function definitions at the same value management study. However this may be too complex and time-consuming, particularly when faced with very large projects. Figure 2.4 summarises the position with regard to function definition. The general rule is that the higher the level of function definition the greater the capacity for changing the project and therefore the greater the potential for improving value. The highest level of function analysis possible should always be selected.

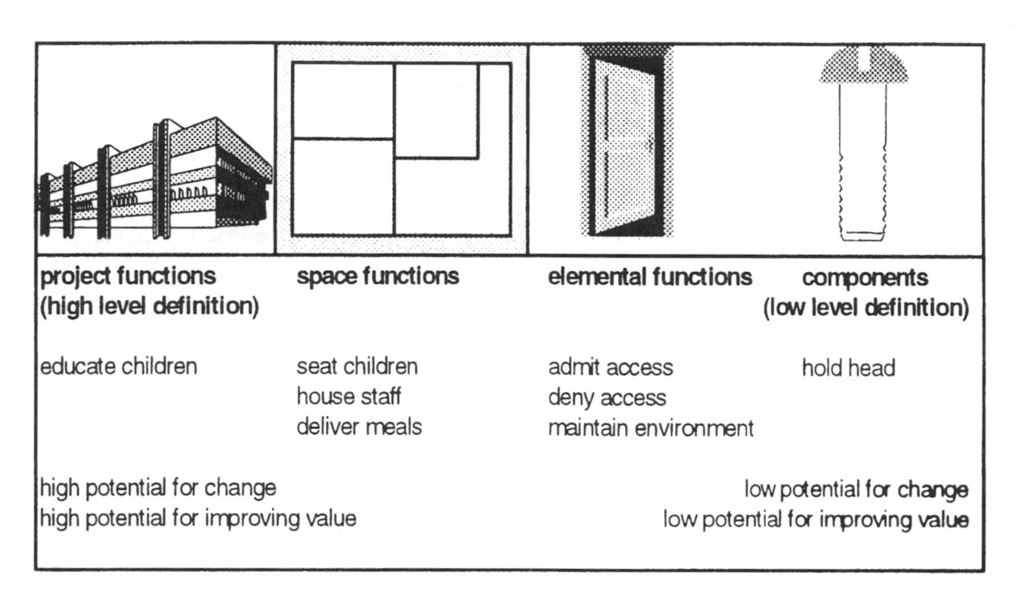

Fig. 2.4 Levels of function definition.

Function before or after design

High level function definition is much more beneficial to the project. It therefore follows that the ideal time to do function definition is before any design has taken place, thereby maximising the potential for change and for improving value. However it is more often the case that some design exists. The problem with this is that the value management team will tend to use this design to define functions. What this in effect means is that they are accepting the functions as designed before they have established that those functions are required. After all, function is based on the requirement of the customer and user, so strictly the functions that come from them will not change, regardless of the presence of an existing design. This may seem like a subtle distinction but it is an important one. Too many design decisions are based on an acceptance of what was done in the past. Often however it is better to start with a fresh approach based on clients' needs only. In that respect only necessary or desired functions will be included.

FAST diagrams

The problem of different levels of function definition was also recognised by the American value managers[7] who tried to show the inter-relationship between these levels through the use of a function analysis system technique or FAST diagram[8] such as figure 2.5. The diagram operates by starting with the primary function or high level function and asking of it 'how is this to be achieved?' The answers to this question form the next layer of functions and the question 'how' is also asked of them until the final function is reached. Operating the opposite way on the diagram and asking 'why' checks that the logic of the diagram is correct. The example shown in figure 2.5 is a crash barrier. The highest level function is to save lives which is achieved by minimising damage, which in turn is achieved by channelling traffic away from the danger, reducing shock in the event of impact and ensuring awareness of the danger. (Only part of the diagram is shown.) Reducing shock in the event of impact is, in turn, achieved by absorbing, transferring and redirecting energy.

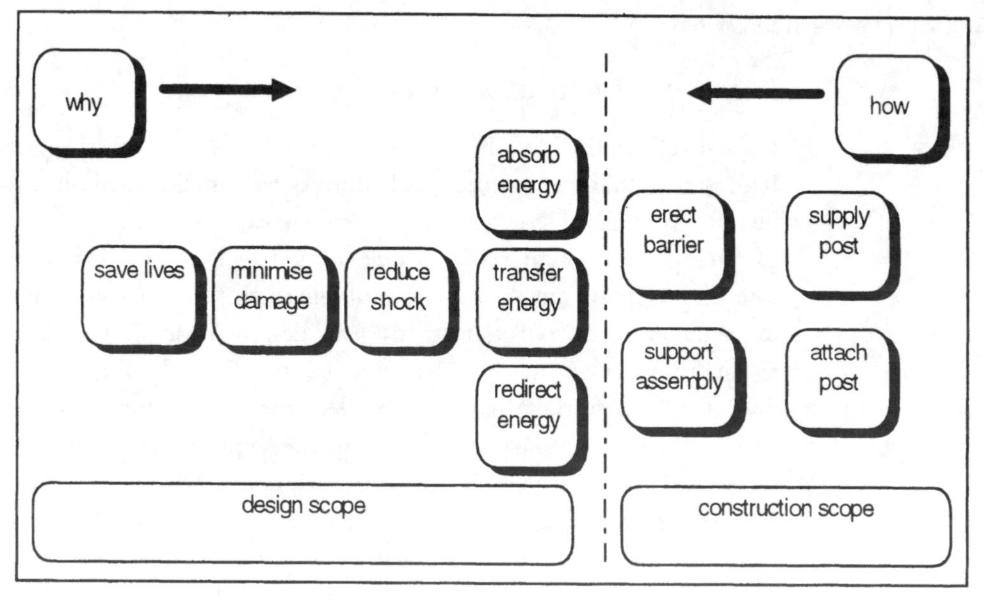

Fig. 2.5 A FAST diagram[9].

FAST diagrams are favoured by the Society of American Value Engineers (SAVE) who claim that they are used extensively. Almost all American textbooks on value management will include examples of FAST diagrams, none of them in our view able to withstand even the mildest of academic scrutiny. It is the conclusion of the authors that the FAST technique, at least in the field of construction, produces little results for what can be a very complex and time-consuming exercise.

Can cost be allocated to function?

Some value management practitioners, on completion of function definition, allocate the estimated cost of the project among the functions[7]. If for example the main functions of a hospital were defined as follows:

- ❑ Treat patients,
- ❑ Diagnose patients,
- ❑ Allow stay,

then the estimated cost of the project is divided up between these functions so that the client can see what it is costing him to provide each of the main functions. This is the method used by Howard Ellegant[10] along with other prominent US value management practitioners. It has some merit in that a client may not want to include a particular function once he becomes aware of the financial implications of providing it. On the negative side however it is very difficult to allocate cost to function, since some items such as the foundations or central plant must be included regardless of whether a function is included or not. The other problem with allocating cost to function is that it really compares apples with pears. We have already seen that function definition should not relate to the existing design: it is best carried out independently. If this is the case then there is no relationship between the function definition and the estimated cost, since the latter is only a reflection of the existing design (figure 2.6). For this reason it is the authors' view that allocating cost to function is largely pointless. However, it is appreciated that there is an argument in favour of it.

Function evaluation

Once functions are defined they are evaluated based on the lowest cost to achieve them. In the example used earlier we considered a window with the sole function of admitting light. The cheapest way

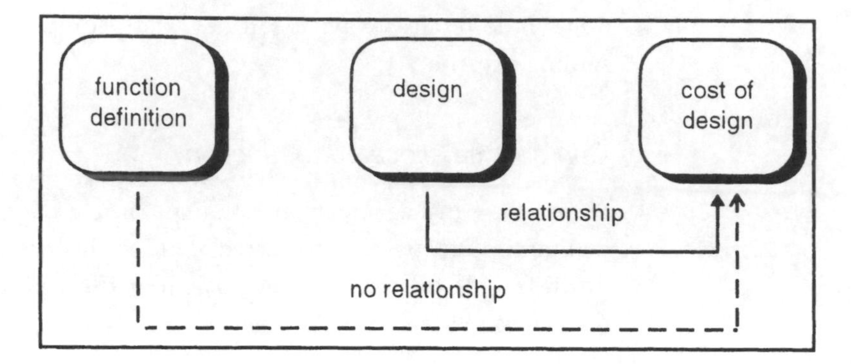

Fig. 2.6 The relationship between function definition and project estimate.

of achieving this function is to put a hole in the wall. This solution is not to be taken literally; it is merely a way of illustrating the value of the function and also a means of generating alternatives that also meet the function. In reality we do not need to accurately price this hole in the wall but just have some idea of its cost. The purpose of function evaluation is a catalyst to creative alternatives. It does not need to be a literal costing exercise.

It is possible therefore that a value management study will produce two sets of costs: the estimated cost of the project allocated between the functions (function costs) and also the lowest cost to achieve those functions. In some value management studies a worth factor is calculated by dividing function cost by the lowest cost to achieve function in order to arrive at a mathematical representation of worth (figure 2.7). These factors can give some indication of the unnecessary cost being spent on functions and may be used to prioritise functions that need to be re-examined. Alternatively no cost is allocated to function and the function is viewed only in relation to the lowest possible cost to achieve it (figure 2.8). These two

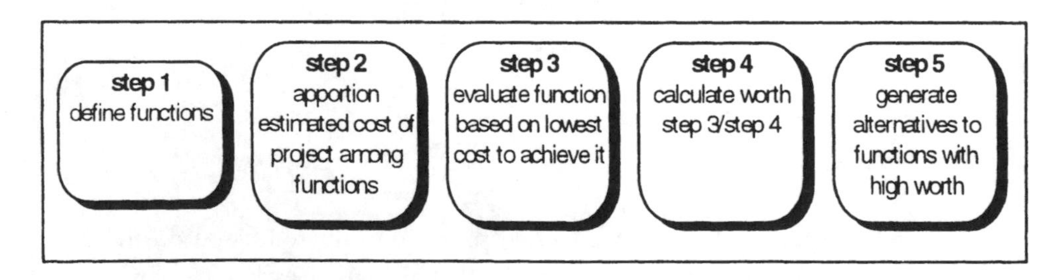

Fig. 2.7 Using estimated costs to evaluate function.

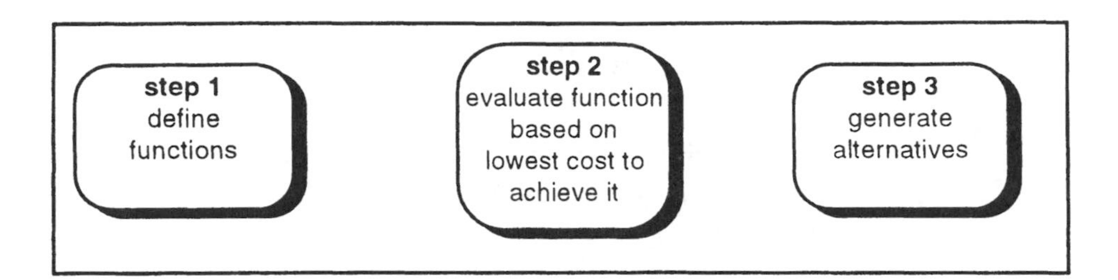

Fig. 2.8 Evaluating functions without the use of estimated costs.

approaches to function evaluation are summarised below. The authors recommend the latter approach.

Alternatives

Once functions have been evaluated the evaluation can be used to generate alternatives by asking 'what else will also achieve the function?'. Going back to the example of the window in the prison cell the cheapest way to 'humanise environment' may be to provide a radio. But what else will also do the job? A TV, a mobile library, bigger cells. All of these may 'humanise environment' and they may do it in a better and cheaper way than a window.

The lowest cost to achieve function acts as a catalyst in generating alternatives that satisfy function but it may still be difficult to get individuals present at the value management study to put forward creative ideas that also meet function. For this reason brainstorming is often used to assist with the generation of alternatives. Brainstorming operates by encouraging wild and outrageous suggestions in the hope that these will generate good and workable ideas, the advantage being that because ridiculous suggestions are encouraged and welcomed there is no embarrassment attached to them or any good ideas that follow. The most stupid idea is viewed as the best one. Participants are encouraged to shout out the ideas as they come to them and formality is discouraged. (Some authors go so far as to suggest alcohol as an aid to this stage of the study[9].) Brainstorming is a general management technique, it is not peculiar to value management. Much has been written about it but we only have enough space in this text to outline its general rules[11]:

- No criticism of ideas is allowed until all ideas have been collected.
- A large quantity of ideas is required.
- All ideas are recorded.
- The best ideas often come from the inexperienced.
- It is a group exercise and ideas should be built on, and used to spark other ideas.

Figure 2.9 summarises function analysis; the core of value management. It is a three-phase approach of function definition, function evaluation and the generation of alternatives. Function analysis is achieved through the verb-noun definition; function evaluation is through the evaluation of the lowest cost to achieve function and alternatives are created through brainstorming.

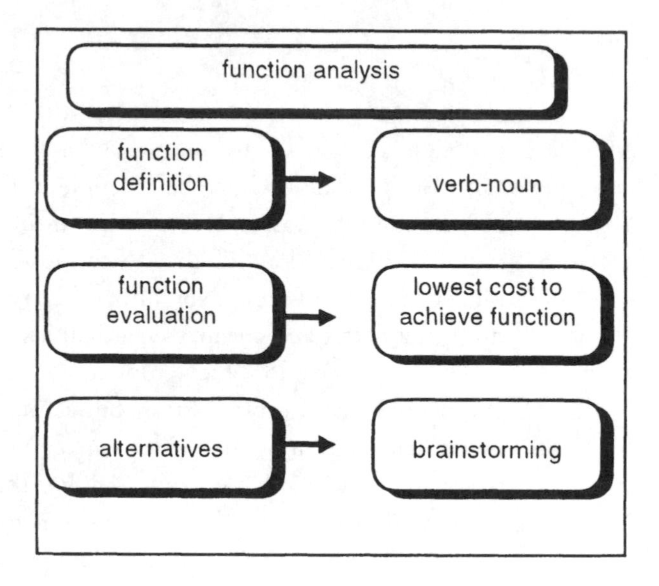

Fig. 2.9 Summary of function analysis.

Function analysis is not easy to implement and may take years of practice to refine; it cannot be learned only from the pages of a book. In addition it is not a stand-alone technique. In order to develop a value management *system* we need not only function analysis but an effective means of carrying it out. How to organise the study, who should do it and when it should take place all need to be carefully considered. The next section of this chapter deals with these additional components of the value management system.

Organisation of the study

An integral part of value management is the job plan which is a five-stage process of organising a study. The five stages are outlined below.

Information stage

At this stage all the information required for the project is gathered.

Analytical phase

At this stage the function analysis takes place in two sections:

- ❑ Define the function.
- ❑ Evaluate the function.

Creative phase

At this stage the function analysis is used as a basis for generating alternatives that meet the functions as defined in the analytical stage.

Judgement phase

At this stage all the possible alternatives that meet the function are evaluated.

Developers' phase

The ideas that are considered worthy of further consideration are selected and developed further.

Who should carry out the study?

The purpose of value management is the optimisation of the needs and wants of those with an interest in the project. It therefore follows that in order to be successful, value management must be carried out as a team exercise. There is really no exception to this and right from the beginnings of value management in the 1940s it has always been viewed as a team approach. Much has been written on group dynamics and leadership in the fields of value management[12] and general management and anyone who wishes to study it in depth will find a large amount of information. However for those interested in the practicalities in relation to value management the following is a summary of the major issues.

Who should constitute the team?

The value management team can constitute the design team or an external team that are new to the project and who have had no previous involvement in it. It may, alternatively, be a mixture of the two, either with or without the presence of the client. In addition specialists may be invited if the project has particular problems and a specialist input is required.

The question of which is the better way of carrying out the study – the design team or an external team – is open to dispute. In the USA the preferred method tends to be an external team, whereas the design team is preferred in the UK. There is no wrong or right answer and it is ultimately the choice of the client. The following is a summary of the advantages and disadvantages of each method.

The advantages of using an external team

- Objectivity.
- The team can be selected for their particular skills whereas the design team is already established.
- Reassurance for the client that the design produced by the design team is a good one.

The disadvantages of using an external team

- The design team may have difficulty accepting the presence of an external team.

- The design team has already formed and overcome many of the teething problems that groups experience. An external team will take time to come together.
- The external team may not really be objective at all. In some cases, particularly in the case of large clients, the external team may want to 'put on a good show'. One way of doing this is to be over critical of the project under study, implying that *they* would have done it better.
- Using an external team is more disruptive than using the design team.
- If function analysis is being used correctly, an external team is not really necessary. The external team may be overly concerned with showing a cost reduction and may reduce specification as a means of achieving it, without any real regard for value or function.
- The external team lack depth in their understanding of the project.
- The external team is expensive.
- The external team raises problems of design liability if they make any changes to the design which later prove defective.

There is of course no such thing as the perfect team. Despite the longer list of disadvantages there are undoubtedly times when an external team may be advantageous – when a project has serious political difficulties for example. However it is the authors' view that in the majority of cases the design team will provide a more effective value management study than an external team.

Who should lead the team?

The value management team leader, or facilitator as they are generally called, needs a skill base that includes an in-depth knowledge of function analysis, group and team building, evaluation of project alternatives and a knowledge of construction. It is unlikely that any member of the design team would have these skills and it will probably be necessary to call on an external facilitator. Value management is a technique built on experience and it is generally recommended than an external facilitator be used.

Should the client attend the study?

A value management study cannot be carried out without the presence of the client, even at the stage of elemental function analysis. The user would also be able to provide a useful insight into the functions of the building and should always be invited. Unfortunately this is rarely the case.

Team organisation

The value management team should not be allowed to become too big. To date, there has been no research into the optimum size of teams for a value management study but anecdotal evidence suggests eight as a maximum.

All members of the value management team should be at approximately the same level of seniority, otherwise junior members tend to feel intimidated. Team members must also have enough authority to make major decisions about the project.

The format of the study

Because of the early and fairly strong influence of the USA on the development of value management in the UK, the technique is often associated with the 40-hour workshop. This system operates by getting the value management team together for a period of 40 hours, away from the normal working environment, so that the study can be carried out free from interruption, thereby providing an environment suitable for the creativity needed to generate alternatives. The idea of the workshop has survived in the UK but rarely is 40 hours deemed necessary and evidence[3] suggests that two days is an average length of study in the UK. The reason for the 40-hour workshop in the USA was largely the efforts of the Department of Defense to standardise the value management procedure and neither SAVE nor any other body suggest it as the optimal solution. American 40-hour workshops are highly cost oriented and the value management team must produce full documentation of all proposals, along with a costing of them. It was the authors' experience that at least half of the time in the 40-hour workshop was used in costing and writing up proposals, making them suitable for presentation to the client. In the UK, where there is much less emphasis on cost, this type of exercise is rarely deemed necessary. The study can therefore be carried out much more quickly. An interesting aside to this is that the US Department of Defense value management programme is so cost-oriented that at the end of each study the savings proposed are divided by the cost of the study to give a return on investment!

As value management is a team exercise it is almost impossible to get away from the concept of the workshop. Some value management practitioners do claim to have an integrated system, value management taking place simply as a part of the normal design procedures. There is no evidence that such an approach could not work but in the opinion of the authors the technique of function analysis is aimed at changing mindset, as such an integrated approach is unlikely to succeed.

Where should the study be carried out?

There is no strict rule about where a value management study should be carried out but most value management consultants or facilitators recommend that it be away from the normal working environment, in a hotel or conference facility.

The timing of the study

A value management study can be carried out at any point in a project's life cycle. The timing should not, in fact, make any difference to the outcome, since the functions of the project do not change simply because the design is at an earlier or later stage of development. However, the amount of redesign that may be required increases as the project develops. As a result the cost to change will also increase, as will the reluctance of the design team to make the changes. Function analysis can be carried out at various different levels (project, space and elemental) and naturally if a function analysis is required on the elements only (possibly because the space allocation is fixed by planning permission or other constraints), it cannot take place until the appropriate level of design is reached.

Some value management writers suggest that the levels of function analysis correspond with the stages of design development and that value management can be carried out more than once[13], to correspond with the design stages outlined below.

Inception

Value management can be used as a means of deciding if the project is really needed. The highest level of function analysis is used. For example a local authority decides to build a new power station. A function analysis shows it is required because the existing power station cannot satisfy demand. The function of the power station is therefore to 'satisfy demand'. This can be achieved either by increasing the supply of electricity or reducing the demand. One way to reduce demand is to encourage people to use energy saving light-bulbs. Rather than build a power station it might therefore be better to give out free energy saving light-bulbs.

Brief

Once it is definitely decided to go ahead with a project, value management can be used to formulate the brief. Space level function analysis is used. Naturally it would still be possible to decide the project was not needed at this stage, but this would result in abortive design. Using value management at this stage, as a means of formulating the brief, is the stage seen as most beneficial by some value management writers and practitioners.[13] For example a ward in an old people's home has functions defined as 'allow stay', 'facilitate nursing', and 'provide food'. The provision of food then generates various alternatives such as self-catering, meals on wheels, restaurant, import from catering or cook on the premises. All these are viable alternatives, particularly given the current emphasis by large organisations on outsourcing.

Outline proposals (35% design)

At this stage the design is more developed and it is possible for the value management team to offer alternatives based on the elemental design and specification. Elemental function definition would be used. Once again value management proposals based on changing the nature of the project or its spatial layout could be put forward but this would require redesign work at this stage.

During the construction of the work

Value management proposals put forward by the contractor have been used for a considerable period, particularly in the USA where these are known as value engineering change proposals (VECPs). Often the contractor is given some financial incentive to make the proposal and this is usually in the form of a percentage of the saving achieved. There are many problems associated with VECPs, not least the question of design liability. In addition many contractors operating under traditional procurement methods particularly feel no incentive to offer change proposals, since the saving achieved may be outweighed by reduction in the scope of work.

How should alternatives be evaluated?

The methods for evaluating value management proposals are numerous but perhaps the one most commonly used is the weighted matrix, an example of which is shown in Table 2.3.

The matrix shows a range of possible light fittings along with some criteria against which they will be evaluated. Each of the criteria is weighted to reflect their importance, so that in the case illustrated in Table 2.3 aesthetics is viewed as the most important criterion, cost as the next most important and so on. Each of the fittings is then given a rating for each of the criteria based on the following scale:

Excellent:	5
Very good:	4
Good:	3
Fair:	2
Poor:	1

In the example shown in Table 2.3, fluorescent fittings, when evaluated against the cost criteria, were given the rating of very good (4) whereas aesthetically they rated poor (1). The rating given is then multiplied by the weighting to give a total cost score of 24 and an

Table 2.3 The weighted matrix.

Method	Initial cost	Maintenance	Aesthetics	Energy impact	Construction time	Total
Weighting	6	3	10	3	3	
Surface mounted incandescent	18/3	3/1	20/2	3/1	6/2	50
Mercury vapour	18/3	9/3	20/2	6/2	6/2	59
Fluorescent fixtures	24/4	12/4	10/1	9/3	9/3	64

aesthetic score of 10. These figures are then collected into an overall total, in this case 64, which can be compared with the other light fittings. The alternative with the highest rating, in this case the fluorescent fittings, is the optimum choice.

Another similar method of evaluation is the SMART system[14]. This is slightly different from the system above in that, as shown in Table 2.4, the allocated weightings must all be a proportion of 1. In addition the rating of each alternative is not based on a predetermined scale but on the degree to which the alternative satisfies the criteria. This score is given out of 100. In the case of the fluorescent fittings, therefore, it might be judged that in terms of cost it satisfies the criteria 80 out of 100 whereas in terms of appearance it only satisfies it 10 out of 100. These scores are then multiplied by the weightings and totalled to give an overall rating. Once again the alternative with the highest score is viewed as the best option.

Other techniques available for evaluation rely less on quantification and more on subjective judgement. Some rely simply on a voting system[15]. The amount of quantitative evaluation that takes place really depends on the stage of project development. At the prebrief stage it may be very difficult to analyse quantitatively the alternatives proposed by the value management team and the choice may simply depend on the preference of the client. If on the other hand the alternatives are well structured then these may be analysed using a weighted matrix.

Table 2.4 The SMART methodology.

Method	Initial cost	Maintenance	Aesthetics	Energy impact	Construction time	Total
Weighting	0.24	0.12	0.4	0.12	0.12	1.00
Fluorescent fixtures	80/19	80/10	10/40	60/7	60/7	83

Value management as a system

Hopefully this chapter has shown that central to the technique of value management is function analysis. Outside of this is a method by which function analysis can be effectively carried out. However because value management is fairly new in the UK and because concepts of value vary, there is as yet no one definitive system; merely a choice of alternative components.

Table 2.5 is a summary of what those components are. A value management system will include an alternative for each component and how those components are put together constitutes the value management system. There is no one correct system and companies should choose the components that best suit them. The summary is not exhaustive. Value management is still in the early stages of development and, as the understanding of the components and their use and interaction develops, the list of alternative components will also expand.

Below are some typical examples of how the components are combined to give value management systems.

The American system

The American system (Table 2.6) is based on a 40-hour workshop carried out by an external team at the 35% design stage. The workshop is structured around the job plan. The practice of American value management uses function analysis loosely based on elements and generated alternatives to them.

Table 2.5 Alternative value management components.

Components	Alternatives
Function definition	Based on project function Based on space function Based on elemental function
Function evaluation	Lowest cost to perform function
FAST diagrams	Use Don't use
Allocate cost to function	Yes No
Calculate worth	Yes No
Generation of alternatives	Brainstorming Other creative techniques
Organisation of the study	Job plan
Group approach	External team Design team Mixture of the two
The value management facilitator	Independent In-house
Format of the value management study	40-hour workshop Two-day study Other as applicable to the project
Location	Outside work environment Within work environment
The timing of the study	Inception Brief Sketch design Construction stage Combination of above Continuous process
Evaluation of alternatives	Weighted matrix e.g. SMART Other mathematical techniques Voting Subjective evaluation

Table 2.6 The American system of value management.

Components	Alternatives
Function definition	Based on elemental function
Function evaluation	Lowest cost to perform function
FAST diagrams	Use
Allocate cost to function	Yes
Calculate worth	Yes
Generation of alternatives	Other creative techniques (ad hoc)
Organisation of the study	Job plan
Group approach	External team
The value management facilitator	Independent
Format of the value management study	40-hour workshop
Location	Outside work environment
The timing of the study	Sketch design
Evaluation of alternatives	Weighted matrix

A case study of value management in the United States

The study

The project was a Department of Defense training building with an estimated cost of $2.4 million. The facilitator was a certified value specialist from a company of value engineering consultants and from a civil engineering background. The team was external comprising an architect, a mechanical engineer, a structural engineer and an electrical engineer. These were selected by the facilitator from consulting organisations. The project was at the 35% design stage. It was single storey reinforced concrete and masonry with pile foundations, concrete floors, built up roof on metal decking. It included a monorail and hoist, sound equipment, compressed air equipment, exhaust systems, fire protection systems, air conditioning and utilities. The total saving achieved was $154 000 out of $535 980 proposed by the value engineering team. The estimated cost of the study was $21 162 therefore giving a 7.3 return on investment. The study was carried out in a 40-hour workshop and produced the following changes to the project.

Value management proposals

- ❑ Reduce amount of acoustical CMU.
- ❑ Delete resilient flooring and seal concrete.
- ❑ Delete suspended ceiling and paint structure.
- ❑ Delete buffer area.
- ❑ Retain exterior insulation.
- ❑ Reduce crane beam span.
- ❑ Change long span roof joists to K series.
- ❑ Use steel framing in lieu of double wall.
- ❑ Use ledger angle in lieu of joists.
- ❑ Revise control joists in exterior walls.
- ❑ Eliminate return air ducts.
- ❑ Add service shut-off valves.
- ❑ Amend air grille detail.
- ❑ Apply demand and diversity factors.
- ❑ Delete MDP panels.
- ❑ Reduce canopy lighting.
- ❑ Reconfigure roof intercom systems.
- ❑ Reconfigure parking lot lighting.
- ❑ Reduce travel lane width.
- ❑ Modify storm drainage.
- ❑ Change type of asphalt paving.
- ❑ Change type of concrete kerb.

The British/European system

A European standard for value management was introduced in 2000, namely EN12973:2000. To quote from the Institute of Value Management (IVM)[16]:

> 'The standard represents a consensus of views and provides a benchmark of best practice. It does not purport to be restrictive or confined to any single commercial sector. It captures the essence of Value Management and provides a framework for its application. It does not advocate the use of any particular techniques but

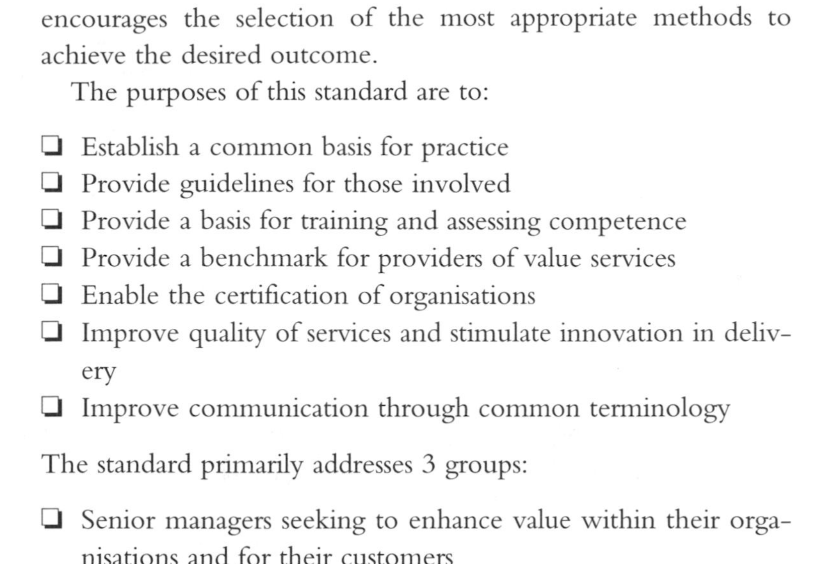

encourages the selection of the most appropriate methods to achieve the desired outcome.

The purposes of this standard are to:

- ❑ Establish a common basis for practice
- ❑ Provide guidelines for those involved
- ❑ Provide a basis for training and assessing competence
- ❑ Provide a benchmark for providers of value services
- ❑ Enable the certification of organisations
- ❑ Improve quality of services and stimulate innovation in delivery
- ❑ Improve communication through common terminology

The standard primarily addresses 3 groups:

- ❑ Senior managers seeking to enhance value within their organisations and for their customers
- ❑ People involved in its application
- ❑ Those involved in training in development'

As stated in the IVM's synopsis of the standard, the standard does not advocate any particular technique or methodology. The approach developed by Kelly[17] is used extensively in the UK (Table 2.7) and typifies good VM practice.

The 'Kelly' system of value management takes place earlier than the American one and is usually carried out by the project design team but led by an external VM facilitator. Function analysis is used to understand objectives and generate alternatives that improve the project.

A case study of value management in the UK[18]

The project was the refurbishment of a public building and the objective of the value management study was the production of a control brief document. The study and the resultant brief covered various sections as outlined below.

Examination of strategic issues

The strategic issues covered by the study included an examination of the users of the building covering both staff and callers. Included was

Table 2.7 A British value management system (based on Kelly[17]).

Components	Alternatives
Function definition	Based on project function and/or Based on space function
Function evaluation	Lowest cost to perform function
FAST Diagrams	Use
Allocate cost to function	Yes
Calculate worth	No
Generation of alternatives	Ad hoc creative techniques
Organisation of the study	Job plan
Group approach	Design team
The value management facilitator	Independent
Format of the value management study	Two-day study
Location	Outside work environment
The timing of the study	Brief Sketch design
Evaluation of alternatives	Weighted matrix e.g. SMART

an analysis of the type of caller, their needs and their queuing behaviour. Other strategic issues covered the political background to the development, the community it serves, the existing building, future potential technology changes, funding, safety and security. The strategic suggestions made by the value management team consisted of:

- ❑ Combine building staff with those currently in another smaller building.
- ❑ Co-ordinate the maintenance and capital budgets to achieve better value.

An analysis of the importance of time, cost and quality

In this section of the study the value management team was asked by the facilitator where the priorities of the project lay in terms of time, cost and quality. The team concluded that time was the least important and that cost and quality ranked equally.

A function analysis

A FAST diagram was used to assist in the function analysis. The FAST diagram started at the highest level of function definition and summarised that the aim of the project was to reflect a corporate approach. This was followed through on the fast diagram to the level of provision of spaces of the building such as toilets and baby-changing area. Unlike the American style FAST diagrams it did not attempt to show the inter-relationship between spaces and elements. The diagram is useful in that it shows how the spaces in the building relate to each other and the higher level functions.

User and users flow diagrams

The value management team produced flow diagrams for public users of the building, security guards, typists, reception areas, interview points, staff generally, finance staff and messenger staff. These diagrams showed how the building would be used by each of the groups and are aimed at improving efficiency of use.

Space definition

This examined the major spaces provided by the project and looked particularly at the quality of the spaces and the environment that they provided. The aim of this was to examine if the quality and the environment was suitable for the building's intended use.

A study of the location and adjacency of spaces

Through use of a matrix and based on the work carried out in the sections above, the value management team produced an adjacency matrix which showed the position of all the spaces in the building in relation to each other. Once again this is aimed at maximising efficiency of use by giving to each space a rating of 0 to 5, where 0 represents no adjacency requirement and 5 represents a definite need for adjacency.

A precontract programme indicating the main action points

This section was largely self-explanatory and was similar to any precontract schedule. The main action points indicated were brief to designers, outline proposals, cost plan, bills of quantities and tender period.

Examination of items requiring immediate action

Based on all the exercises outlined above, the value management team produced a list of items requiring immediate action. There were 21 items in all including:

- ❑ Check the need for a basement fire escape.
- ❑ Resolve underground drainage.
- ❑ Establish floor loading.
- ❑ Carry out security survey.
- ❑ Reassess size of waiting area and conference room.
- ❑ Reassess number of reception points, interview rooms and interview booths.

The agenda of the workshop

The study was carried out over two days to the following agenda:

Day one

- ❑ Introduction to value management
- ❑ Facilitators present information
- ❑ Study of strategic issues
- ❑ Lunch
- ❑ Time, cost and quality study
- ❑ Time schedule
- ❑ Function analysis

Day two

- ❑ Review of the previous day
- ❑ User flow diagrams
- ❑ Lunch
- ❑ Functional space definition
- ❑ Quality and environment study

- Adjacency
- Review and close

The value management team

The team comprised the following people:

- Four members of the client body which included three quantity surveyors
- Seven members of staff or building users
- Two architects from the architectural consultants
- The mechanical and electrical engineer
- Two facilitators

The British case study highlights how much value management has developed in the UK. Although function analysis is still a core part of the study, the workshop also includes a much more global examination of the project's objectives and user needs. In addition, unlike the American study there is little emphasis on cost and emphasis is on improving the use of the building. Unfortunately the example shown above is still quite rare. However, value management in Britain is at the cross-roads of its development and it is possible that this type of study will soon become commonplace. After all, the above case study was carried out over two days and with the exception of the facilitators, used only those already involved in the project. As such, costs would be small, yet as the study shows, the benefit to the project can be considerable.

Other British case studies can be found in the work by Green and Popper[19].

Value management in Australia

Because of the similarities in procurement approaches between the UK and Australia, value management in Australia is similar in character to value management in the UK. It could however be argued that the practice of value management in Australia is more firmly entrenched than in the UK. As stated in Chapter 1, the Department of Public Works and Services (DPWS), New South Wales Australia has, since the mid-1990s, made a value management study a mandatory requirement on all majors projects, i.e. projects over Aus.

$5 million. DPWS has produced an extensive set of *Value Management Guidelines*[20] which covers:

- the nature of value management
- value management policy
- concept and application
- process
- procedures
- evaluation matrices
- case studies

In our view, DPWS is an excellent example of a public sector organisation using value management to ensure best value for the client. The fact that DPWS has successfully implemented value management for a number of years is testament to the fact that, as far as DPWS is concerned, value management is not a passing fad but a productive management tool. The DPWS *Value Management Guidelines* has interesting case study material and is part of the DPWS Total Asset Manual series which contains a wealth of information on topics such as life cycle costing and risk management in addition to value management. The DPWS also has a website[21]

The Japanese system

Unlike the American and British systems, Japanese VM is not a one-off exercise but a continuous process carried out under the umbrella of the construction project[22]. The Japanese view VM more as a philosophy than a system, one that operates at all stages of the construction cycle, including planning, maintenance and environmental protection (see Table 2.8).

A case study of value management in Japan

In Japan the procurement systems for public and private work are significantly different and this naturally has an effect on the way value management is carried out. In the private sector design and build is the usual form of procurement and under these circumstances value management is carried out by the construction company. In the

Table 2.8 A Japanese value management system.

Components	Alternatives
Function definition	Based on project function Based on space function
Function evaluation	Lowest cost to perform function
FAST Diagrams	Use
Allocate cost to function	No
Calculate worth	No
Generation of alternatives	Other creative techniques
Organisation of the study	Job plan
Group approach	Design team
The value management facilitator	In-house
Format of the value management study	Other as applicable to the project
Location	Within work environment
The timing of the study	Continuous process
Evaluation of alternatives	Subjective evaluation

public sector procurement is more akin to the traditional form of procurement where design and construction are separated. In this case value management is carried out in two stages. First by the in-house designers or consultants and based on the design as it progresses, and second by the contractor. The value management becomes part of the procurement and a typical process is illustrated below. This is taken from the Kobe City Housing Company project – a multiple function hall with an audio-visual hall and housing combined in a 17 storey building with basement. It has a total of 14,261 m^2 floor area and total cost of 41 billion Yen. The value management procedure was as follows:

- *Step 1*
  Designate contractors who can take part in the tender.
- *Step 2*
  Give the contractors the design carried out by the designers, the costing and the procedures for value management.

- *Step 3*
  Contractors make value management proposals based on the existing design.
- *Step 4*
  The value management proposals are examined by the client organisation and classified as good or bad.
- *Step 5*
  The contractors choose some of the proposals which were judged good and make a final proposal to the client which is submitted with a total cost.
- *Step 6*
  The contractor is selected based on the lowest price which must be the same as, or below, the original price at the time of contractor selection.

Some typical proposals from the above project were as follows:

- Change the frame above the fifth floor from *in situ* to precast concrete.
- Change the formwork below the fifth floor from timber to metal to facilitate reuse.
- Use recycled hardcore in lieu of new.
- Change the position of the roof ducting.
- Change the pipe layout of the sprinkler system.

The above is by no means a standard value management system, as in Japan there are many methods of implementing value management. In some cases but by no means all a percentage of the saving is given to the contracting company which proposes it.

Why are the systems different?

The value management systems have developed separately because the construction industries and business cultures within which they exist are different[23]. A study of these cultures leads to a better understanding of the systems that operate within them. In the context of this book it also shows that when importing construction management techniques from other countries, consideration must be given to the effect of business culture. Investigation of these differences provides insight for the organisation wishing to develop bespoke value management systems that suit their own business culture. The influence of business culture on value management cannot be overstated and the following illustrates how it affected the development of value management in the USA and in Japan.

Differences in the style of management

In the information-gathering process American managers place greatest emphasis on action in the immediate period and are satisfied with fast results. The Japanese manager on the other hand places greatest emphasis on ideas and looks to the future, obtaining most satisfaction from a range of possibilities as opposed to a definite solution. This indicates that in the USA a fixed system of VM that shows a definite output may be desirable, whereas in Japan a more evolving process may be more appropriate. This is in fact reflected in the two VM systems: the American model provides a relatively short, definite process that produces a tangible set of results, whereas the Japanese model is continuous and less obvious.

In terms of information evaluation the American manager places emphasis on logic and is most satisfied with solutions. The Japanese manager on the other hand emphasises human interaction at this stage, largely based on past experience. Satisfaction is obtained not from a solution but a more holistic assessment of the problem. In terms of the evaluation of ideas produced by VM it means that in the USA a mathematical form of evaluation may be preferred, whereas in Japan a more intuitive system may be appropriate.

Differences in management systems

Styles of management lead to differences in management systems[24]. The Japanese and Western management systems are different and this has a direct relationship on the operation of value management. Table 2.9 summarises the main differences in management systems.

Examining the systems of value management in different countries it can be seen how they have been influenced by management systems. With both British and American VM the single most problematic area has undoubtedly been the human relationships and this is largely a result of the management system being depen-

Table 2.9 Differences in management systems.

Japanese management depends on	Western management depends on
The group[25]	The individual
Long-term orientation	Short-term results
Ambiguity of responsibility	Clearly defined roles and tasks
Control by humanism	Controlled by logic and reason
Employment of people	Employment of function
Harmony	Seen to be fair
The acceptance of unequal relationships	Equal relationships
Constant improvement	Results
Respect for age and seniority	Less important in business relationships
Decision-making by consensus	Decisions by seniors
Control by face-saving	Control by superiors
Task control	Ideological control

dent on the individual. Unlike in Japan, group behaviour is not inherent and is therefore problematic. The need for short-term results has also dictated that the American system is very results oriented, with most studies calculating a return index that shows the saving achieved in relation to the cost of the VM study. This invariably leads to VM teams trying to find the greatest savings possible, often at the cost of overlooking function analysis. American management's emphasis on logic and reason has resulted in an approach to VM that is much more definite than the Japanese approach, which although a clear system is integrated and less well defined. In addition any business culture that is based on long-term planning will probably be more receptive to the concept of value as opposed to cost. This is also true of a system controlled by ideology as opposed to one controlled by task.

The relationship between value management and quantity surveying

This chapter should have illustrated that by its very nature value management has little relationship to the traditional role of the quantity surveyor. This is not to say that quantity surveyors are incapable of broadening the scope of their traditional service to encompass value management: they are, and some quantity surveying practices already offer value management services. However, development of value management has not been in the QS field because the QS is better disposed to the role of the value management facilitator than any of the other professions: it is simply because the Royal Institution of Chartered Surveyors have made positive efforts to encourage its development.

Conclusion

The concept of evaluating functions as a means of assessing value is problematic for two reasons. First, value is subjective and second, value changes with time.

A piece of jewellery such as a wedding ring has fairly limited functions: it may be to show that the wearer is married, or it may be purely for decorative purposes. The value to the owner, though, particularly if they have owned it for many years, may be much greater than indicated by the function because it has sentimental value. Most of us acknowledge that there is such a thing as sentimental value and most of us recognise that this value, although real, cannot be defined or quantified.

In addition to its subjective nature the value of items may vary over time. As outlined at the beginning of this chapter, the value of a battery-operated radio may drastically increase when a snowstorm cuts the electricity supply. In a less dramatic example the value of some goods, such as children's football kits, may diminish simply because a newer version is brought on to the market.

We have therefore no real measure of value other than an assessment of the value to an individual or group of individuals at some given point in time. What we do have, however, and what we can always achieve, is a definition and evaluation of the functions that we either need or want an item to perform. This distinction between needs and wants is an important one. In value management a function does not only apply to what the project needs to do but also to what we would like it to do. If only needs were considered all buildings would be square with flat roofs and linoleum floors. Clearly though, many clients wish to provide prestige and comfort in their buildings and these can be equally defined as functions.

The big problem with function analysis in the field of construction is that it is difficult to decide whose functions and values to assess. The value to the government who support the construction of a new hospital may be increased political support in that area. The value to the health authority may be an increased and better facility for their patients and staff. The value to the staff may be an improved working environment and the value to the local people may be a better facility in their area that may reduce delays and travelling time. In addition to this is the value to society, which is the provision of better infrastructure that improves the quality of life for all. The question that arises is which of these subjective and changing values should be assessed? For the purpose of value management the answer is all of them.

But is it really possible to make an assessment of the needs and wants of all those involved in a hospital project, from the politicians down to the patients? The answer in the context of value management is that we translate the needs and wants of those with an interest in the building into functions: that is, what the users need and want the building to do. This is not an easy exercise and skill is required in teasing out from the multitude of conflicting interests the true functions of the building and then providing the most economical design solution that meets those functions.

Although function analysis is the central pivot of value management, other components are required for function analysis to operate effectively. Within these other components there are alternatives and the choice of alternatives which constitute the value management system is dependent on many factors such as the project, the time scale, the design team and the business culture of the industry. It therefore follows that no one system of value management is correct, as the most effective system will vary with the project. Flexibility therefore is a key to good value management.

Understanding the component parts of value management and the best time to use them is vital for successful value management. However value management is a fairly new technique and information on when to use what is not always available and may not be for many years to come. More practice of value management is needed, along with more research, before these questions can be answered.

One of the biggest mistakes that is made with value management is to confuse it with cost reduction. We hope this chapter has illustrated that value management has in fact very little to do with cost: it is a design process. The fact that cost reduction often comes about as a result of value management is more a consequence of it than an objective.

References

1. *Chambers English Dictionary* (1989) 7th edn. W&R Chambers Ltd and Cambridge University Press.
2. Fong, P.S.W. (1996) VE in construction: a survey of clients' attitudes in Hong Kong. *Proceedings of the Society of American Value Engineers International Conference*, Vol. 31.0.
3. Palmer, A.C. (1992) *An investigative analysis of value engineering in the United States construction industry and its relationship to British cost control procedures*. PhD thesis, Loughborough University of Technology.
4. Miles, L. (1967) *Techniques of Value Analysis and Engineering*. McGraw-Hill Book Company, New York.
5. Mudge, A. (1971) *Value Engineering, a Systematic Approach*. McGraw-Hill Book Company, New York.
6. Heller, E.D. (1971) *Value Management, Value Engineering and Cost Reduction*. Addison-Wesley, Reading, Mass.
7. Snodgrass, T. & Kasi, M. (1986) *Function Analysis: The Stepping Stone to Good Value*. Department of Engineering Professional Development, University of Wisconsin.
8. Bytheway, C.W. (1965) Basic function determination technique. *Proceedings of the Society of American Value Engineers Conference*, April, Vol. 2, 21–3.
9. Kelly, W. (1986) *You and Value What Not*. VEST, Walla Walla Washington State, USA.
10. Ellegant, H. (1990) Value engineering before you draw! A better way to start project. *Seminar Notes: The Value Management of Projects*. Heriot Watt University, Edinburgh, UK, 24 April.
11. Dell'Isola, Al. (1988) *Value Engineering in the Construction Industry*. Smith Hinchman & Grylls, Washington DC.
12. Barlow, D. (1989) *Successful interdisciplinary ad hoc creative teams*. Appleseed Associates, Ohio.

13. Kelly, J.R. & Male, S.P. (1991) *The Practice of Value Management: Enhancing Value or Cost Cutting*. Department of Building, Engineering and Surveying, Heriot Watt University, Edinburgh, Scotland.
14. Green, S. (1994) Beyond value management: SMART value management for building projects. *International Journal of Project Management*, **12** (1), 49–56.
15. J.J. Kaufman Associates, 120006 Indian Wells Drive, Houston, Texas, 77066.
16. Institute of Value Management website (2002) http://www.ivm.org.uk/vm_europe_standard.htm.
17. Kelly, J. & Male, S. (1993) *Value Management in Design and Construction – the Economic Management of Projects*. E & FN Spon, London.
18. A VM study facilitated by John Kelly, Senior Lecturer, Department of Building, Engineering and Surveying, Heriot Watt University, Edinburgh, Scotland.
19. Green, S.D. & Popper, P.A. (1990) Value engineering: the search for unnecessary cost. *The Chartered Institute of Building*, Occasional Paper no. 39, May.
20. New South Wales Department of Public Works and Services (2001) *Value Management Guidelines*. Report no. 01054. Government of New South Wales, Sydney.
21. Department of Public Works and Services website (2002) http://www.game.nsw.gov.au.
22. Society of Japanese Value Engineers. (1994) *Management of Construction in a Changing Period*, (translated from Japanese) Society of Japanese Value Engineers, Tokyo.
23. Palmer, A. (1996) A comparative analysis of value management systems in the UK, USA and Japan. *CIB International Conference, Construction Education and Modernisation*. Beijing, on CD-ROM.
24. Baba, Keiso. (1990) Principal nature of management in Japanese construction industry. *Journal of Construction Engineering and Management*, **116** (2), 351–63.
25. Walker, A. & Flanagan, R. (1991) *Property and Construction in Asia Pacific*. Blackwell Science, Oxford.

Chapter 3
Constructability

Introduction

The terms 'constructability' and 'buildability' will not be found in any standard dictionary. They are terms which are specific to the construction industry and have meaning only to those operating within the confines of the industry. It would be fair to say that although the principles underpinning these terms are gaining more and more acceptance in a number of countries, the use of the words 'constructability' and 'buildability' is not yet commonplace in the vocabulary of many construction industry practitioners.

In the context of this chapter the terms are taken to be synonymous and can be used interchangeably. In the interests of consistency we have opted for 'constructability' in preference to 'buildability', except when referencing or quoting from authors who have chosen the alternative. We have avoided attaching any subtle difference between the terms, ignoring, for example, the folklore that buildability is British and constructability is American or that constructability encompasses wider system boundaries than buildability.

The industry-specific nature of constructability makes it unique in comparison to all other concepts covered in this book. Concepts such as total quality management and reengineering straddle a range of industries whereas constructability can make the unusual claim that it is the only management concept in the past 30 years to have been designed and developed by the construction industry for the construction industry.

Origins

By comparison with other industries the separation of the processes of design and construction is unique to the construction industry. This compartmentalisation of functions has been highlighted over the years in reports such as the Simon Report[1], the Emmerson Report[2] and the Banwell Report[3]. In response to this perceived deficiency, the Construction Industry Research and Information Association (CIRIA)[4] in 1983 focused attention on the concept of 'buildability'. The view taken by CIRIA was that buildability problems existed

> 'probably because of the comparative isolation of many designers from the practical construction process. The shortcomings as seen by the builders were not the personal shortcomings of particular people, but of the separation of the design and construction functions which has characterised the UK building industry over the last century or so.'

CIRIA defined buildability as 'the extent to which the design of the building facilitates ease of construction, subject to the overall requirements for the completed building'. The CIRIA definition focused only on the link between design and construction and implied that factors which are *solely* within the influence or control of the design team are those which have a significant impact on the ease of construction of a project.

About the same time in the USA the Construction Industry Institute (CII) was founded with the specific aim of improving the cost effectiveness, total quality management and international competitiveness of the construction industry in the USA[5]. Constructability was, and still is, a significant component of the CII's research and development work. The CII definition of constructability is wider in scope than the CIRIA approach and defines constructability as

> 'a system for achieving optimum integration of construction knowledge and experience in planning, engineering, procurement and field operations in the building process and balancing the various project and environmental constraints to achieve overall project objectives.'[6]

More recently in the early 1990s in Australia, the Construction Industry Institute, Australia (CIIA) has tailored and developed the CII constructability process to Australian conditions. In doing so the CIIA has amended the CII definition of constructability to:

> 'A system for achieving optimum integration of construction knowledge in the building process and balancing the various project and environmental constraints to achieve maximisation of project goals and building performance.'[7]

The goals of constructability

The goals of constructability are determined by the scope which constructability is intended to cover. The 1983 CIRIA definition limited the scope of the concept to the relationship between design and construction. This is illustrated in figure 3.1.

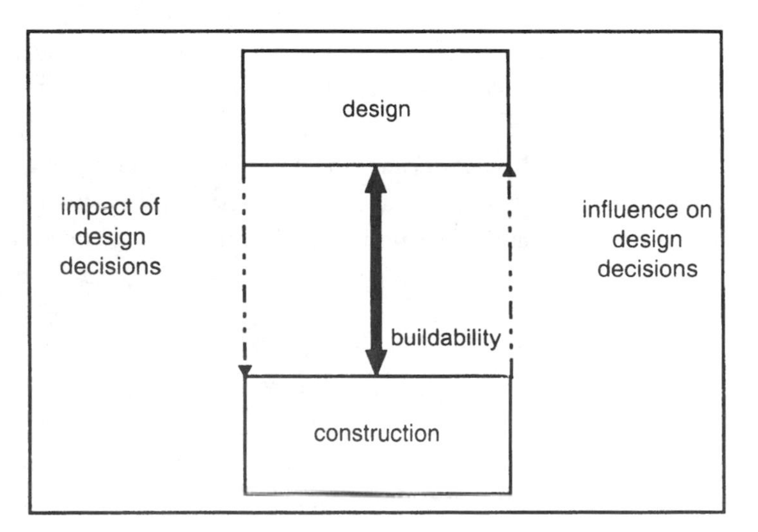

Fig. 3.1 The scope of 'Buildability' as defined by CIRIA.

In systems terms the delineation of the scope of a conceptual model is known as the system boundaries. The system boundaries of the CIRIA model are quite narrow, viewing constructability purely as a design-oriented activity. Griffith[8] takes the view that this approach is in fact highly restricted and has resulted in the loss of some momentum in the uptake of constructability in the UK. Several researchers have discussed the difficulty in arriving at the appropriate boundaries for the constructability model[8,9,10]. On the one hand, if the boundaries are too wide, there is the inherent danger of applying a simplistic approach which equates constructability with a set of motherhood statements which have very little prospect for practical implementation. Conversely, if a very narrowly focused approach is taken then this may fail to realise the full potential of the concept.

A workable concept of constructability needs to recognise that there are many factors in a project environment which impact on the design and construction processes, and the link between design and construction and the maintenance of the building. This can be illustrated diagrammatically in figure 3.2[11].

Figure 3.2 demonstrates the factors influencing the design process, the construction process, and the quality and performance of the finished product. Only when the complex interaction of these factors is acknowledged can the potential of constructability be achieved. Of prime importance is the acceptance of the view that buildability does not equate simply to the ease of construction but is also concerned with the appropriateness of the finished product. This can be seen in one definition which defined buildability as[12]

> 'the extent to which decisions made through the whole building procurement process , in response to factors influencing the project and other project goals, ultimately facilitate the ease of construction and *the quality of the completed project*'.

(This definition has been adopted by the New South Wales Government, Construction Policy Steering Committee[13].) Moreover, attention to constructability does not cease with the completion of the building. The constructability of maintenance activities, in, for example, the installation, removal and replacement of materials, finishes, services and equipment is equally as important throughout the life of the building as the constructability of the initial construction phase.

Decisions which are made upstream of the design stage can impose constraints on the design decision process. At the same time, decisions which may not have been made by the designer concerning intermediate functions between design and construction such as docu-

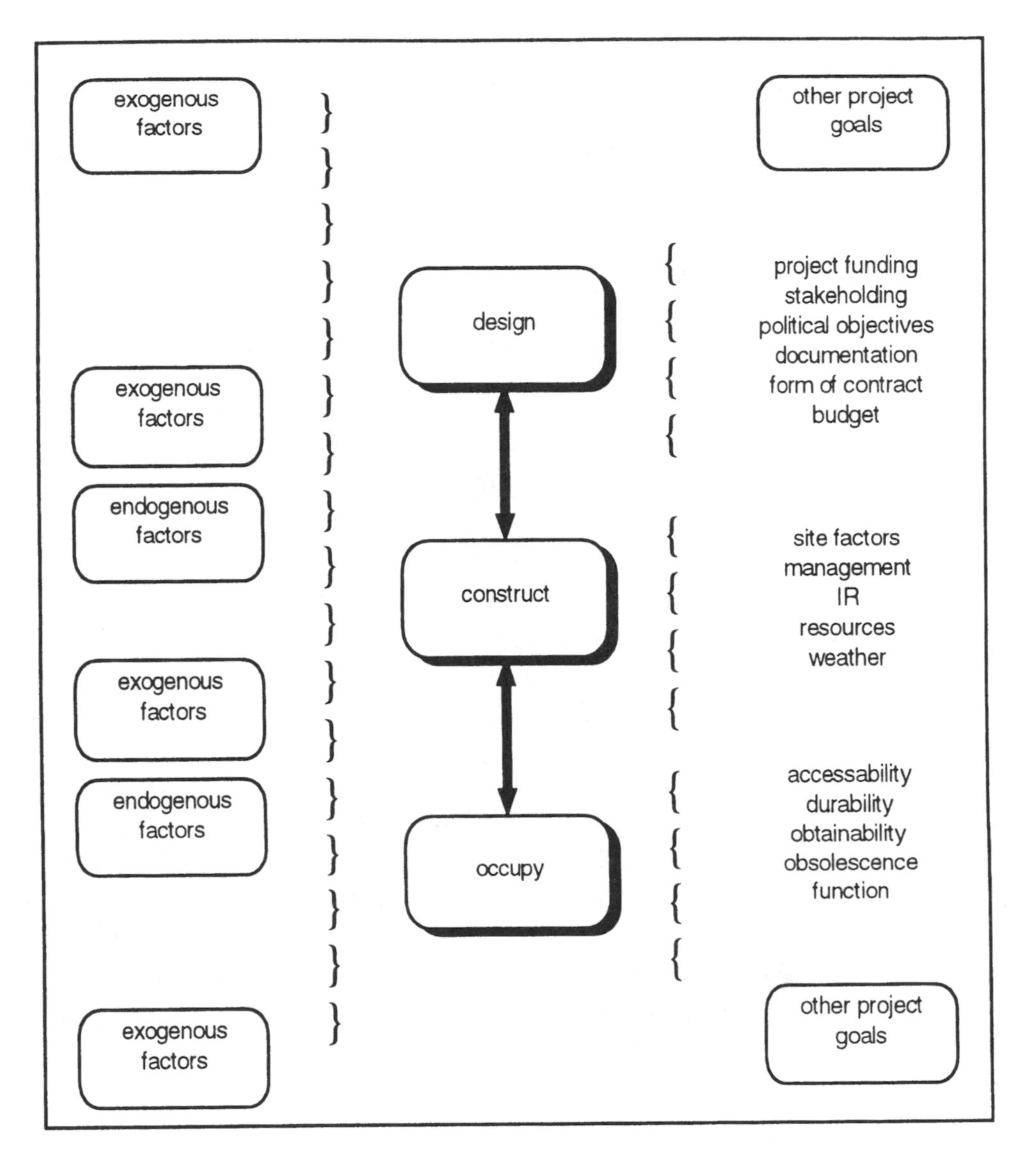

Fig. 3.2 Performance framework for managing constructability.

mentation, contractor selection, choice of contract form procedures and so on, may have a significant impact on the construction process. Similarly the maintenance manager who is conventionally regarded as being downstream of the design and construction decision-making process will be significantly affected by upstream buildability decisions. Referring to figure 3.2 it can be seen that each stage of the project, from design through to occupancy, is influenced by forces such as exogenous factors, endogenous factors and project specific goals. (An exogenous factor is one which originates from, or is caused by external forces. An endogenous factor is one which originates from, or grows from within.) An approach which takes into account the influence of exogenous and endogenous factors, together with project goals, must of necessity deal with issues of some complexity. If the complex nature of the procurement process is not acknowledged there is the danger that constructability can lead to common denominator strategies which are of little practical use.

Implementing constructability

Constructability principles

The previous section stressed the need to take a balanced view and establish appropriate system boundaries if the goals of constructability are to be achieved. This question of balance has been addressed by CII Australia[7] who, in conjunction with the CII[6], have produced a best-practice, how-to-do-it constructability manual (for a detailed explanation of the CIIA Constructability Files in particular and constructability principles in general, readers are referred to Griffith and Sidwell[5]). The CII primer and the CIIA constructability files[1] are not just comprehensive checklists for industry practitioners but are also in keeping with the conceptual model of constructability illustrated in figure 3.2. The contents of the CIIA Constructability Principles File are as follows:

- Implementation advice on how organisations can establish a constructability programme.
- Flowcharting indicating the applicability of the principles of constructability at the various stages of the project life cycle.
- Executive summaries of the principles of constructability.
- Twelve principles of constructability.
- Database to record examples of savings from constructability.

The CIIA advocates a structured approach which identifies the following five stages in the procurement process:

- feasibility
- conceptual design
- detailed design
- construction
- post construction

The 12 principles are then mapped on to the procurement process. These principles are as follows:

(1) **Integration** – (constructability) must be made an integral part of the project plan.
(2) **Construction knowledge** – project planning must actively involve construction knowledge and experience.
(3) **Team skills** – the experience, skills and composition of the project team must be appropriate for the project.
(4) **Corporate objectives** – (constructability) is enhanced when the project teams gain an understanding of the clients' corporate and project objectives.
(5) **Available resources** – the technology of the design solution must be matched with the skills and resources available.
(6) **External factors** – external factors can affect the cost and/or programme of the project.
(7) **Programme** – the overall programme for the project must be realistic, construction sensitive and have the commitment of the project team.
(8) **Construction methodology** – project design must consider construction methodology.
(9) **Accessibility** – (constructability) will be enhanced if the construction accessibility is considered in the design and construction stages of the project.
(10) **Specifications** – project (constructability) is enhanced when construction efficiency is considered in the specification of the development.
(11) **Construction innovation** – the use of innovative techniques during construction will enhance (constructability).
(12) **Feedback** – (constructability) can be enhanced on similar future projects if a post-construction analysis is undertaken by the project team.

Table 3.1 illustrates the distribution of the CIIA 12 principles over

Table 3.1 The distribution of the 12 principles over the procurement process.

Feasibility	Conceptual design	Detailed design	Construction	Post-construction
[1]	[1]	1	1	1
2	[2]	2	2	4
[3]	[3]	[3]	(7)	[12]
[4]	[4]	(5)	(9)	
[5]	(5)	6	[11]	
[6]	(6)	(7)		
7	[7]	[8]		
8	[8]	[9]		
	[9]	[10]		

the five stages of the procurement process. The principles are plotted on a three-point scale of importance depending on their location in the procurement stages. For example 'external factors' is of very high importance [6] at the feasibility stage, (6) is highly important at the conceptual design stage but is of lesser importance 6 at the detailed design stage, and has no influence on the construction and post-construction stages. These levels are by way of guidelines to users of the CII principles method and are not unchanging. They are simply intended to help the user determine the principles which are likely to be of relevance and of significance at particular stages in the project life cycle.

The various participants in the project process will have different roles and responsibilities in terms of the 12 principles and will have different responsibilities at different stages of the project's life cycle. The decisions taken by the participants need to be co-ordinated in order to optimise the constructability performance of the project. Otherwise, individual participants may take different strategies towards achieving goals within their sphere of influence, thus compromising the overall constructability performance of the project.

Figure 3.3[14] proposes a constructability implementation planning framework which identifies and co-ordinates the decision roles and responsibilities of individual project participants throughout a project's life cycle. This enables constructability plans to be developed for individual projects, which allow individual participants to identify not only their own roles and responsibilities but also allows them to see what other participants are or should be doing at every stage of

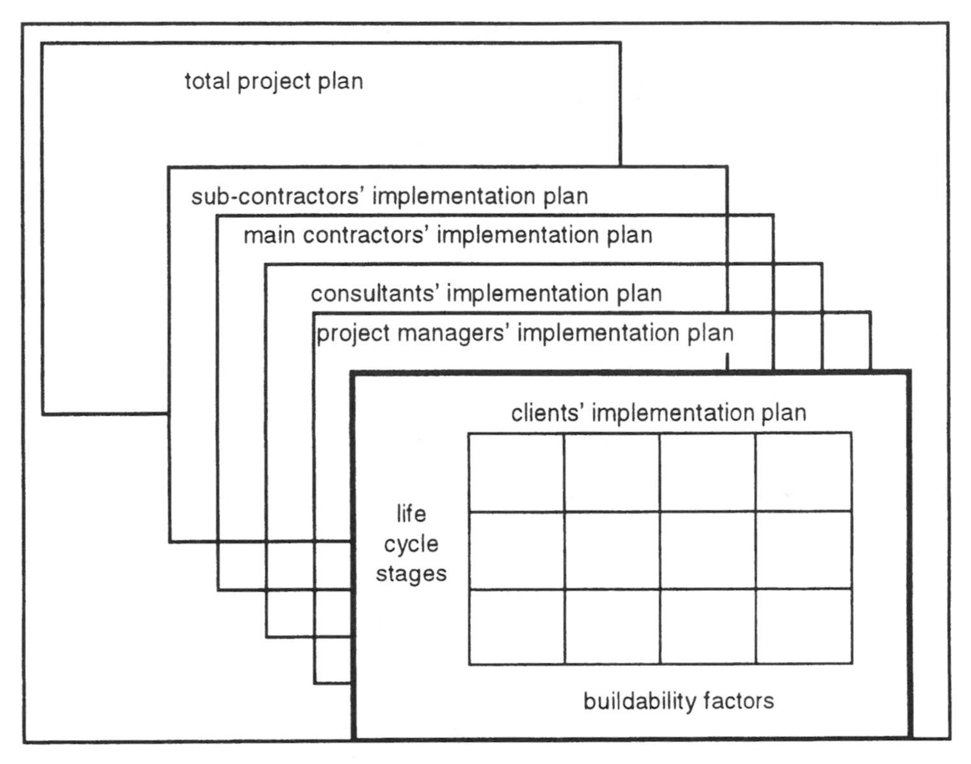

Fig. 3.3 Constructability implementation planning framework.

the project life cycle and therefore allows better integration and co-ordination.

The essence of this approach is that constructability can be enhanced by individual participants exploiting construction knowledge to maximise opportunities and develop best options to meet project objectives in a co-ordinated way and also by adopting collective review processes such as value management.

Constructability in practice

To successfully implement constructability management, the client or the client's representative should in the first instance put in place a programme which clearly specifies the primary project objectives and allows constructability to be assessed as a project performance attribute. Constructability objectives should also be clearly identified for the different roles and responsibilities of the various members of the project team. This can be done through a performance specification relating to the time-cost, cost and quality criteria addressing the needs of the client and the users, and using the framework (illustrated in figure 3.3) which co-ordinates the consideration of constructability principles by all the project team members throughout the whole project life cycle.

Although setting a mechanism in place which co-ordinates the constructability principles among the project team members is an important aspect of implementing constructability, it is equally important to recognise the significance of the timing of the input by the various team members (figure 3.4).

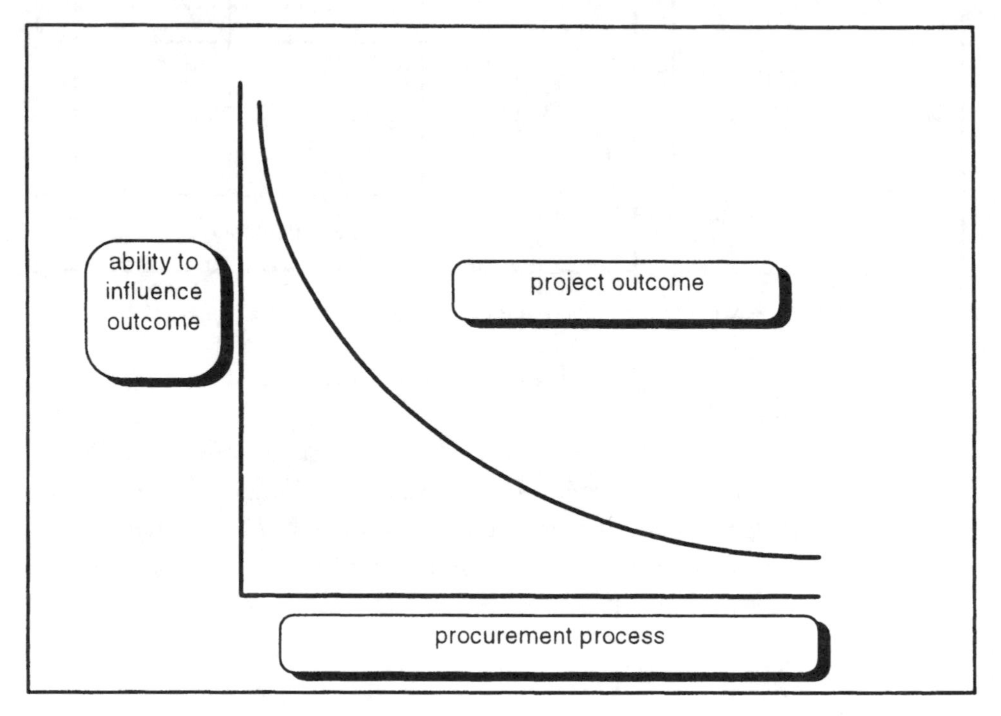

Fig. 3.4 Cost influence/Pareto curve.

The importance of timing is illustrated by the Pareto principle which contends that decisions taken at the early stages in the project life cycle have greater potential to influence the final outcome of the project than those taken in the later stages.

'During the early phases such as conceptual planning and design, the influence of decisions on project cost is high but diminishes rapidly as the project moves into the construction stage. It is vital therefore, to make decisions leading to various design and construction aspects of the project from the first instance. Early contractor involvement in the project is seen to be the desirable approach to attain good buildability'[15].

This is a good juncture to consider the relationship of the design and build form of procurement to constructability (design and construct in Australia). There is no doubt that the design and build form of procurement does provide for early contractor input into the decision-making process. Given that CIRIA's initial interest in buildability stemmed from the view that 'buildability problems existed, probably because of the comparative isolation of many designers from the practical construction process', it may then be thought that the adoption of design and build would go a long way to solving the problem of the dislocation of the process of design from the process of construction. To some extent this is true, and most commentators subscribe to the view that design and build does indeed provide a project environment that is more conducive to 'good' constructability than traditional procurement methods such as selective tendering[5]. However, the adoption of design and build, or derivatives such as management contacting and construction management, do not automatically result in a better approach to constructability.

The key to the successful implementation of constructability is in having effective communications between members of the project team, and while a design and build form of procurement can streamline the lines of communication, the team members themselves must be committed to the constructability concept if a successful outcome is to be achieved. There is one view[16] that the reason for the occurrence of 'bad' constructability is partly due to changes in emphasis in the education of modern designers and that there is insufficient emphasis given to building construction in the curriculum of architecture courses. An alternative view would be that there is not enough emphasis given to management, or conversely it could be argued that not enough emphasis is given to design appreciation in construction management courses. The permutations are endless, and will always be coloured by the professional background of the commentator. The nub of the issue is that until building professionals receive a multi-disciplinary education, there will always be barriers to achieving 'good' constructability.

Constructability and the building product

The principal goal of constructability is to produce the best product i.e. a building, making the best possible use of resources. Given this focus, it is surprising that so little attention has been given to the relationship between constructability and the building in-use. This is despite the fact that decisions which have a significant impact on the ease of access to components for maintenance and replacement, as well as the ease of assembly and disassembly, are made in the early stages of the project life cycle. Research work[17] has explored the relationship between decisions taken at the inception stage of the procurement process and the downstream effects on the building in-use. This can be loosely summarised as follows:

Buildings are durable assets which are often threatened with technical and functional obsolescence long before the end of their structural life expectancies. The maintenance and replacement of building components is a common everyday occurrence in the life cycle of a building as a response to physical deterioration, technological obsolescence, changes in performance and functional criteria. Building performance should not therefore be judged at a specific stage in a building's life cycle but should be considered over the life cycle of a building as whole[18]. The potential for a building to extend its useful life is an important factor in its life cycle performance. The operational efficiency of a building through its complete life cycle is determined largely by the characteristics of its original design, the construction or assembly processes and the demands generated by operational requirements, maintenance, alterations and ultimately disassembly or demolition[19]. The level of a building's performance is largely reflected in the quality of decisions taken in the early stages of the project[20].

Efficient building maintenance and renewal is characterised by the need for easy access to, and disassembly and repair or replacement of existing building components. Planning for better performance with respect to these demands should be viewed as being within the domain of constructability. Constructability is often regarded as purely a design issue relating to the ease of initial building assembly. This approach limits and indeed diverts early project decision-making away from the consideration of the performance of the building throughout its full life cycle.

The key to improving building performance is in effective information management, particularly at the early project stages where decisions have the greatest potential to influence project outcomes. The quality of decision-making is promoted by the identification of critical issues and the availability of timely and relevant information.

In terms of constructability-oriented maintenance and renewal management, valuable decision support for project policy-makers, designers and other stakeholders can be provided by the systematic identification and tracking of important decision situations and planning accordingly to meet required performance objectives. An integrated project decision support framework which can facilitate the implementation of a constructability-oriented maintenance and renewal management strategy is illustrated in figure 3.5.

Conceptually, the framework integrates the following functions:

- action planning (constructability-oriented action plan)
- cumulative decision recording (integrated project description)
- information access, filtering and processing
- access to relevant decision tools and evaluation techniques
- communication and co-ordination among project participants.

The constructability-oriented action plan will identify issues having a major impact on constructability and alert relevant decision-makers to the issues they need to deal with. The timing of these actions is governed by the development of the integrated project description through each stage of the project life cycle. This project description is the representation of the project in terms of all the decisions which have been made. At the feasibility stage, the project description could be a statement of all perceived project objectives which eventually develops into a design brief. At various stages of the project life cycle, aspects of the project description may be manifested in forms such as sketch design drawings, working drawings, specifi-

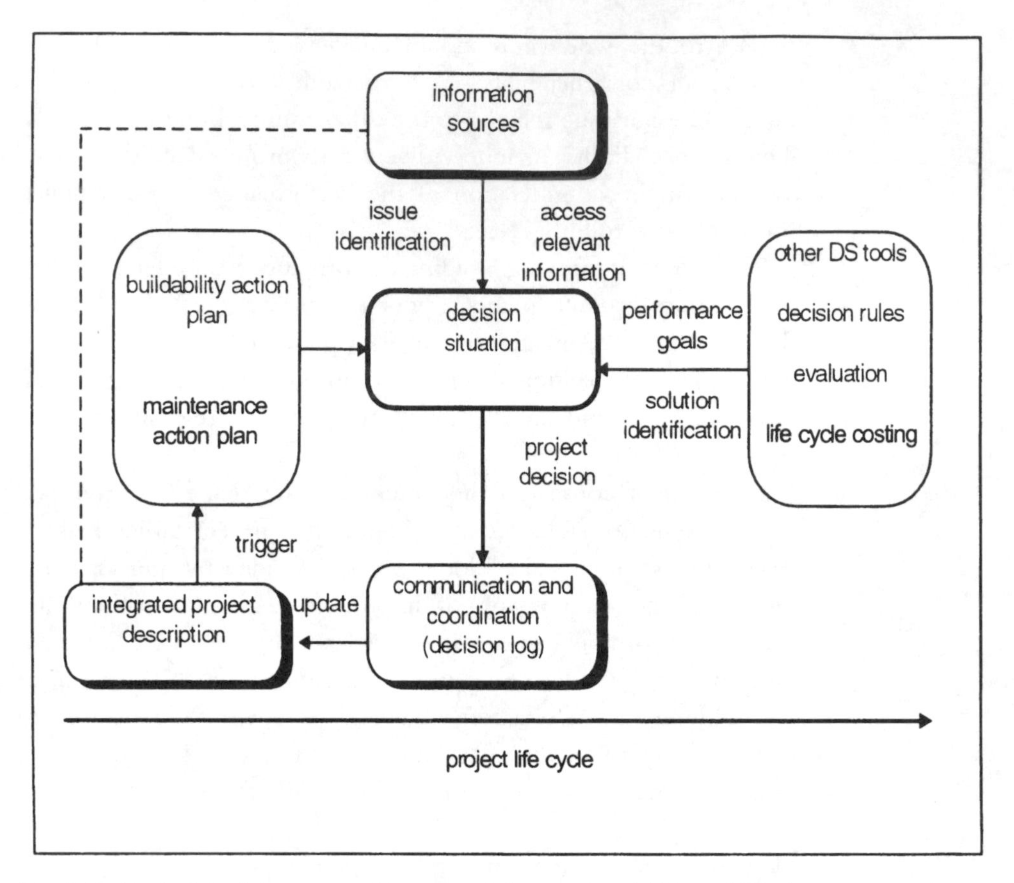

Fig. 3.5 Project decision support framework for constructability-oriented management.

cations, bills of quantities, and shop drawings. In practice, the compartmentalisation of the different functions in the project process tends to fragment the project descriptions put together by different participants. This contributes to communication and co-ordination problems. The use of computerised tools for integrated applications and networking would allow the development of an integrated project description by which all project participants are informed. The quality of responses to the decision situations generated by the action plan will determine the eventual performance of the project and is enhanced by access to relevant information, and expertise and tools to formulate, evaluate and select optimal solutions As each decision situation is acted upon, the integrated project description is continually updated, keeping all participants informed on the status of the project and also triggering the progress of the action plan to the following decision situation.

Consideration of maintenance implications at the design stage is not a new idea[21]. However, success in this area has been moderate. Maintenance and renewal consideration in the early project stages has been impeded for a variety of reasons, the most critical including time pressures, unaware decision-makers and ill-informed designers.

Time is critical in the early stages of design and numerous conflicting issues are competing for the attention of the designer. The decision-maker invariably has little time to spend on investigating simple yet important operational or servicing issues, let alone more far-reaching concepts regarding ease of replacement and renewal considerations. Maintenance management and renewal considerations should be addressed at a policy level by the client or the project manager. Maintenance-oriented design then becomes one of the objectives of the project and maintenance management and renewal becomes an integral part of project management[22]. If maintenance management is established explicitly in the early phases of the project when project objectives are being defined, then resources will be allocated to the examination and investigation of the future maintenance of the building. This implies that particular expertise in the various areas of technical servicing, replacement and component disassembly and reassembly may be retained to advise on the assessment of design alternatives.

Historically the attention to maintenance and renewal performance has been limited because designers responsible for new facilities have been unaware of the post-construction problems and inefficiencies experienced by those who have to maintain and operate them and which their decisions impact upon[19]. Unfortunately designers have limited access to the buildings they design in the post-construction period and therefore tend to be divorced from the maintenance problems that flow from poor design.

Decision-makers if aware of possible issues relating to maintenance and renewal have rarely had access to the right information that would aid project decision-making. Project decisions made by designers have drawn upon information from three areas: personal knowledge and experience, formal reference sources and project description at that point[23]. In the building industry there has been a tendency for practitioners to rely on personal knowledge, their own or that of peers[24]. While this has been effective on numerous occasions it has serious limitations[23]. Easy access to formal reference material that is relevant, properly indexed and written in a usable form would provide a reliable base from which designers could make informed decisions.

Decision-makers need to be given the required decision support in terms of access to relevant information, expertise and tools. The model illustrated in figure 3.5[25] demonstrates that if issues regarding the operation/maintenance/renewal phase are considered, then such decision support tools as life cycle costing, maintenance records, post-occupancy evaluations and asset registers could be accessed. A systematic approach ensures that the requirements generated by different stages of a project life cycle and the inputs of many compartmentalised project decision processes can be integrated. The overall constructability-oriented management strategy then provides the logical continuity for a single integrated framework that does not overtake the functions of individual project decision-makers but serves as a common decision support system to all the project participants. The proposed framework allows the application of information technology (IT) to address the communication, co-ordination and information management problems that confront most project decision-making processes.

Ease of access and assembly for building maintenance and renewal are significant issues in the overall performance of buildings over their complete life cycles. A constructability-oriented strategy can provide a logical vehicle for improving maintenance and renewal performance. This is achieved by providing relevant decision support to the early project stages, where the decisions have the greatest impact on overall building performance. The proposed information management framework provides a singular decision support mechanism for all project decision-makers and takes into account the dynamic requirements of the project across its complete life cycle. The harnessing of information technology enables the complex co-ordination and communication demands to be met to overcome problems of compartmentalisation.

[Note: The above description is based on on-going research being undertaken by the Building Performance Research Group at the University of Newcastle, Australia into the identification of significant factors in constructability-oriented management.]

Good and bad constructability

The point has now been made repeatedly that constructability is a dynamic attribute which is project specific. While it is important that examples of constructability, both good and bad, are recorded and a constructability knowledge base is developed and expanded, it is also important to recognise the limitations in terms of the transferability of good constructability features from project to project. In the past, particularly in the UK, many of the recorded case studies seemed to have concentrated on what had gone wrong, i.e. examples of bad constructability. Although these may have a salutary effect, greater improvements are likely to be achieved by concentrating on positive aspects. This is why the CII and the CIIA Principles Files[6,7] are an important step forward because by creating a clear conceptual framework the principles underlying examples of good constructability can be identified and *transferred* to new project situations.

The development of a constructability index based on indicators of success which concentrate on what went right in a project, rather than what went wrong, has been proposed[10]. The underlying assumption is that the greatest gains are likely to be achieved in the management of constructability information and in recording the decision-making process during the procurement cycle. This is not to dispute the importance of improvements in construction technology, but to view constructability as a management driven rather than a technologically driven process. In support of this view the Building Research Establishment estimates that 90% of building design errors arise because of failure to apply existing knowledge, strengthening the argument that one of the most important aspects of constructability is not the lack of information but rather the lack of management of information. Of course, gains can and have been made in the technology areas. Examples, particularly in the UK, include the use of modular co-ordination, design rationalisation, standardisation, precasting and dry finishes. However, without the proper tools for managing information for the project decision-making process, and without a clear conceptual model from which to operate, it is unlikely that the full potential of the constructability approach can be exploited.

As outlined above, much of the previous efforts to produce a constructability index or scale has been derived from case studies which seem to concentrate on what went wrong with projects. A more productive way forward is for constructability targets to be identified as focal points for decision-making.

Essentially constructability can be considered as a project attribute which is:

- within the influence of those who shape the project process
- measurable against indicators of success (reflecting the ease of construction and quality of the completed project).

From this working definition, constructability can be represented as having three dimensions. These are:

- the participants (stakeholders and decision-makers)
- the constructability factors
- the stages in the building procurement process.

Figure 3.6 represents a three-dimensional conceptual model of constructability. The three dimensions define the boundary locations of possible solution spaces where improvements in constructability may be achieved. The three-dimensional model provides a framework to locate the main factors which need to be considered for each project situation to improve constructability.

The participants

The participants comprise two groups. The first group is the stakeholders. These are persons who have some interest in the possible outcomes of the project process. The second group consists of persons who make decisions which influence these outcomes. Some of these possible outcomes are directly or indirectly associated with the constructability of a project.

Constructability factors

The factors (which roughly equate to principles in the CII principles approach) are those which one or more 'participants' can influence or

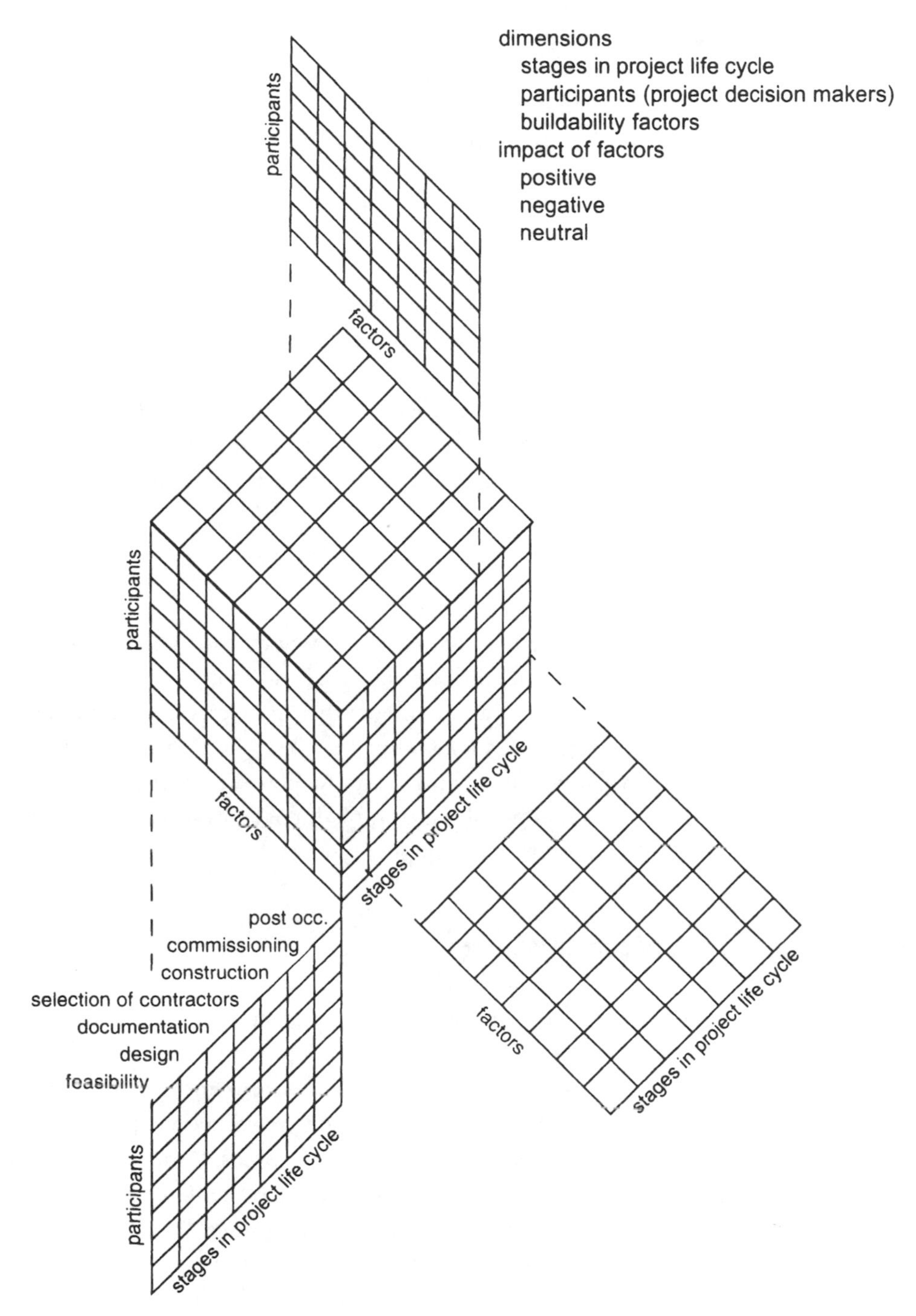

Fig. 3.6 A conceptual model of constructability.

respond to in terms of their decisions, and which would impact on the ease of construction or the quality of the completed project. These constructability factors may be exogenous factors, endogenous factors or project goals (other than the goal of constructability) (see also figure 3.2).

Stages in the project life cycle

Constructability is not limited to the design-construction subsystem. Decisions which impact on constructability include those which are made upstream of the design stage of the project. At the same time, the standards by which constructability is to be assessed include the performance of the project after construction.

Application

The approach recognises that in some cases a factor which is positive in terms of impact on constructability in one project may be neutral or even negative in impact on another project. Within imposed constraints, project decisions may be taken so as to minimise or neutralise negative effects, and maximise positive effects. (This work is still at a developmental stage and is linked to research of a more general nature on decision support systems[25].)

Achieving the benefits of constructability

The benefits of constructability can occur at all stages of the procurement process, although the Pareto principle dictates that the earlier in the process that constructability thinking is incorporated, the greater will be the impact and the greater will be the potential for time and cost savings and quality improvements. Some claim[12] that the implementation of constructability management can lead to significant quantifiable improvements in project performance in terms of time, cost and quality. In addition to the quantifiable measures, constructability management can also lead to qualitative improvements in the project process as well as the building product. They cite benefits such as:

- better project team work
- improved industrial relations
- better forward planning
- higher productivity and smoother site operations.

A major study in 1988 which investigated the application of constructability in a project management concluded that[15]:

> 'The contribution of construction personnel to the design of the projects was significant.
>
> The iterative relationship between construction and design in various project phases led to tangible benefits ranging from cost and time savings and ease of construction, to the elimination of union demarcation and other industrial relations problems.
>
> The rationalisation of the design, which involves simplification, modularisation and repetition of design detailing, is essential to the achievement of constructability.
>
> The achievement of constructability is influenced by technical factors such as building technology/systems, project planning and scheduling, etc. in the building process.
>
> There are many other factors, particularly non-technical factors, associated with the management of building projects (such as project delivery systems, communication, quality of management) which need to be considered as part of the process of achieving constructability.'

The CII[6], CIIA[7], and others[5,16,26] have all documented cases studies and detailed design situations which demonstrate the benefits of constructability . For readers who wish to explore constructability at a detailed level we would recommend Adams[16] and Ferguson[26] who give numerous examples of design detailing and site organisation situations where constructability principles have been applied. This former work lists 16 design principles and gives examples covering all of these. These design principles are as follows:

(1) Investigate thoroughly.
(2) Consider access at the design stage.
(3) Consider storage at the design stage.
(4) Design for minimum time below ground.
(5) Design for early enclosure.
(6) Use suitable materials.
(7) Design for the skills available.
(8) Design for simple assembly.
(9) Plan for maximum repetition/standardisation.
(10) Maximise the use of plant.
(11) Allow for sensible tolerances.
(12) Allow a practical sequence of operations.
(13) Avoid return visits by trades.
(14) Plan to avoid damage to work by subsequent operations.
(15) Design for safe construction.
(16) Communicate clearly.

Other sources give a big picture of the benefits of constructability. Case studies documented by the CIIA[27] typically demonstrate the performance achieved through the systematic application of constructability management. Two examples of these, the Toyota Car Manufacturing Facility at Altona, Victoria, and the Australis Media Centre at the Technology Park in South Australia, give a flavour of the benefits of applying constructability in the following situations:

The Toyota Car Manufacturing Facility comprises a building area of 120 000 square metres and was completed in 107 weeks (three weeks ahead of schedule), at a cost of Aus. $161.2 (Aus. $18.8 million below budget) to a high standard of quality and safety. One of the significant actions taken in the project was to set up a major roofing and cladding prototype to test and establish the joint and opening details to achieve the required air tightness and weather performance required for the paint shop facilities.

The Australis Media Centre required the construction of a new 7300 square metre facility and was completed on a very tight schedule of nine months at a cost of Aus. $12 million, with cost savings which allowed additional scope of work within the budget. The constructability management strategy employed ensured close coordination and liaison between the project team members and early resolution of design, detailing and construction method considerations.

Quantifying the benefits of constructability

Measuring and quantifying the benefits of constructability is not a straightforward matter, particularly if constructability is applied in the broad sense discussed in this chapter. There are several reasons why this is so. In the first instance many of the benefits of constructability, such as better teamwork, are qualitative rather than quantitative. In the second instance there is often a synergistic, or knock-on, effect in terms of constructability actions. In other words, the whole is always greater than the sum of the parts: thus simply measuring the time, cost and quality improvements of individual aspects of constructability and summating these will not guarantee that the overall impact of constructability has been captured. By way of example if representatives of a local planning authority are included in the project team (as was the case in the Australis Media Centre case study[27]) then this may result in fast-tracking the approval process, which might in turn have very fundamental implications for downstream activities. The third reason why the benefits of constructability are difficult to quantify is a methodological one. The ideal way to measure the benefits of constructability would be to conduct parallel case studies of identical projects comparing 'constructability projects' with 'non-constructability' projects. This approach is clearly highly improbable and impracticable and, at best, could only be applied to a very limited sample. It is possible to find instances where very senior executives of major companies are prepared to declare publicly that 'they have trialed constructability on three major projects and have saved 5% on cost and 13% on time'[27]. While not disputing the veracity of these claims, the methodology used in quantification is open to challenge on the basis that the comparison relies on a *hypothetical* comparison which assumes an alternative situation where constructability was not used at all, or was used badly.

There have been some attempts to relate managerial actions and building project performance in terms of time, cost and quality:[15,28,29,30,31,32]

> '...the practice of constructability analysis is more common in Australia than in the United States or the United Kingdom ... an average of 18 man months is spent on buildings of 20–30 storeys in height, compared with an average of four man-months on similar US projects, as part of the design process.'

Ireland[30], in what Hon *et al.*[15] describe as a first attempt to seriously quantify constructability, concludes that:

> 'The construction time of major projects varies by a factor of almost four to one; the cost implications of the time variation are that savings in holding charges and gain in rental for the quickly constructed projects, compared with the industry average time, can be as large as 50% of the building cost.'

Although it is clear from such studies and also from experience in the USA that the order of magnitude of the benefits of constructability can be significant, a method of quantifying the indicators of success is not yet a reality.

Conclusion

It may be thought that the virtues of constructability are self-evident and that the principles of constructability are indistinguishable from the principles of good multi-disciplinary team working. This is a reasonable assumption, and one which is difficult to dispute. Constructability is about managing the deployment of resources to their optimum effect. To do so means establishing seamless communication between members of the team. This, in turn, means the breaking down of traditional barriers and altering professional mindsets. Builders must be empathetic to the views of architects and vice versa. Clients must be prepared to play their part in responsible decision-making. All members of the project team must be prepared to play a proactive role and address the complete building cycle from inception through to occupation. Expressed in this light we can begin to see why constructability is not yet a commonplace construction management tool. A saying by Jortberg[5] ruefully describes the current situation as 'designers and engineers don't know what they don't know'. Constructability has come a long way from the early CIRIA definition that it is 'the extent to which the design of a building facilitates ease of construction, subject to the overall requirements of the completed building', but it still has a long way to go to attain the goals of the CIIA definition of constructability as being 'A system for achieving optimum integration of construction knowledge in the building process and balancing the various project and environmental constraints to achieve maximisation of project goals and building performance.'

References

1. Central Council for Works and Buildings. (1944) *The Placing and Management of Building Contracts.* HMSO, London.
2. Emmerson, Sir H. (1962) *Survey of Problems before the Construction Industries.* HMSO, London.
3. Committee on the placing and management of building contracts. (1964) *Report of the Committee on the Placing and Management of Building Contracts.* HMSO, London.
4. Construction Industry Research and Information Association (CIRIA). (1983) *Buildability: An Assessment*, Special Publication 26. CIRIA Publications, London.
5. Griffith, A. & Sidwell, A.C. (1995) *Constructability in Building and Engineering Projects.* Macmillan, London.
6. Construction Industry Institute (CII). (1986) *Constructability: a Primer.* CII University of Texas, Austin, Publication 3-1.
7. Construction Industry Institute Australia (CIIA). (1992) *Constructability Principles File.* CIIA University of South Australia, Adelaide.
8. Griffith, A. (1986) Concept of buildability. *Proceedings of the IABSE Workshop 1986: Organisation of the Design Process.* Zurich, Switzerland, May.
9. Bishop, D. (1985) *Buildability: the Criteria for Assessment.* CIOB, Ascot, Berks.
10. McGeorge, D., Chen, S.E. & Oswald, M.J. (1992) The development of a conceptual model of buildability which identifies

user satisfaction as a major objective. *Proceedings of CIB International Symposium*, Rotterdam, May.

11. Chen, S.E. & McGeorge, D. (1993/94) A systems approach to managing buildability. *Australian Institute of Building Papers*, **5**.
12. Chen, S.E., McGeorge, D. & Varnam, B.I. (1991) *Report to the Government Architect, New South Wales, Buildability Stage 1.* TUNRA, University of Newcastle.
13. New South Wales Government, Construction Steering Committee. (1993) *Capital Project Procurement Manual.* NSW Govt., Australia.
14. Chen, S.E., McGeorge, D., Sidwell, A.C. & Francis, V.E. (1996) A performance framework for managing buildability. CIB-ASTN-ASO-RILEX *Proceedings of the International Symposium: Applications of the Performance Concept in Building*, 9–12 December, Tel Aviv, Israel.
15. Hon, S.L., Gairns, D.A. & Wilson, O.D. (1988) Buildability: A review of research and practice. *Australian Institute of Building Papers*, **39** (3).
16. Adams, S. (1989) *Practical Buildability*. Butterworth, London.
17. Chen, S.E., London, K.A. & McGeorge, D. (1994) Extending buildability decision support to improve building maintenance and renewal performance. In: *Strategies and Technologies for Maintenance and Modernisation of Building, CIB W70 Tokyo Symposium, 2 (1191–1198)*. International Council for Building Research Studies and Documentation CIB, Tokyo, Japan.
18. Powell, J.A. & Brandon, P.S. (1990/91) Editorial conjecture concerning building design, quality, cost and profit. *Quality and Profit in Building Design*, E. & F.N. Spon, pp. 3–27. Quoted in Mathur, K. & McGeorge, D. Towards the achievement of total building quality in the building process. *Australian Institute of Building Papers*, **4**.
19. Bromilow, F.J. (1982) Recent research and development in terotechnology: Building maintainability and efficiency research and practice. *CSIRO National Committee of Rationalised Building*, Australia, 3–13.
20. Speight, B.A. (1976) Maintenance in relation to design. *The Chartered Surveyor*, October 1968, quoted in Seeley (1976).
21. Seeley, I.H. (1976) *Building Maintenance*. Macmillan, London. 1976.

22. Kooren, J. (1987) *Where Does Maintenance Management Begin?* Building Maintenance Economics and Management, 245–56, E & FN Spon, London.
23. Leslie, H. (1982) *An Information and Decision Support System for the Australian Building Industry*. CSIRO National Committee of Rationalised Building, Australia.
24. Mackinder, M. & Marvin, M. (1982) Design decision in architectural practice: the roles of information, experience and other influences during the design process. *Research Paper No. 19*, Institute for Advanced Architectural Studies, University of York. Quoted in Leslie (1982).
25. McGeorge, D., Chen, S.E. & Ostwald, M.J. (1995) The application of computer generated graphics in organisational decision support systems operating in a real time mode. *Proceedings of COBRA 1995 RICS Construction and Building Research Conference*. Edinburgh.
26. Ferguson, I. (1989) *Buildability in Practice*. Mitchell, London.
27. Sidwell, A.C. & Mehrtens, V.M. (1996) Case studies in constructability implementation. *Construction Industry Institute Australia*, Research Report 3.
28. Ireland, V. (1983) *The role of managerial actions in the cost, time and quality performance of high-rise commercial building projects*. Unpublished PhD thesis, University of Sydney, Australia.
29. Ireland, V. (1984) Managing the building process. Report: New South Wales Institute of Technology, Sydney.
30. Ireland, V. (1985) The role of managerial actions in the cost, time and quality performance of high-rise commercial building projects. *Construction Management and Economics*, **3**, 59–87.
31. Ireland, V. (1986) An investigation of US building performance. Report: New South Wales Institute of Technology, Sydney.
32. Sidwell, A.C. & Ireland, V. (1987) An international comparison of construction management. *Australian Institute of Building Papers*, **2**, 3–11.

Chapter 4
Benchmarking

Introduction

For decades athletes have looked outside their own sports for new techniques that have formed the basis of very successful training programmes. Emil Zatopek, who was the only man ever to win three long distance athletic gold medals at one Olympic games, learned his techniques from the Army[1]. Other athletes such as Ron Hill, one of the world's greatest marathon runners, used the carbohydrate loading diet invented by Swedish physiologists to improve his performance[2]. Others are said to have lived on a diet that included turtle blood and ground rhinoceros horn.

What is common in all these experiences is that individuals looked outside the scope of their own sports or disciplines to find ways of improving. They were using the training methods already accepted in their own sports but these were not enough. Everybody was using them. In order to really succeed they needed something else. They needed a competitive edge.

Parallels for this can be seen in industry. For example when Henry Ford II was faced with rescuing a failing business he took new concepts of management from his competitor, General Motors[3]. However although there are examples like Ford, industries and companies are generally reluctant to look beyond their own sphere in order to find the competitive edge. For reasons of competitive fear, lack of resources or simple conservatism, organisations tend to rely on the tried and trusted methods that exist within their own limited spheres. This is not to say that these tried and trusted methods are worthless.

The message therefore is that the search for superiority is a three-layered pyramid of success. In the case of the athlete it means they first must do the best they can. Second they must do the best that others in their field can, by studying the training methods of other athletes. Finally they must do the best there is by looking to the outside world and examining techniques in the fields of physiology, psychology and nutrition and apply these to their own training programmes.

This idea also applies to the management of a company. In order to gain a competitive edge a company needs to look at itself first. It needs to examine its own systems and methods of working and make the necessary improvements. It also needs to look at its own industry to learn the best methods from it and try to achieve those best practices itself. Finally it needs to look outside its own industry to learn the best methods from other industries and to try to achieve those best practices also.

This process of looking outside one's own sphere, be it to other divisions, companies or industries, is basically one of comparison. It would be pointless investigating other companies if the information gathered was not used as a standard against which to measure one's own performance. Once this comparison is made, however, and a performance gap established, it can be used as a basis for setting goals aimed at the improvement of one's own practices.

This pyramid of success based on the comparison with others is the basis of benchmarking (figure 4.1). Benchmarking is the comparison of practices either between different departments within the company, or with other companies in the same industry, or finally with other industries. The aim of benchmarking is to achieve superiority.

Definition of benchmarking

What makes benchmarking different from other management techniques is the element of comparison, particularly with the external environment. However, benchmarking is more than simple comparison and before defining it formally it is worth considering its other essential ingredients.

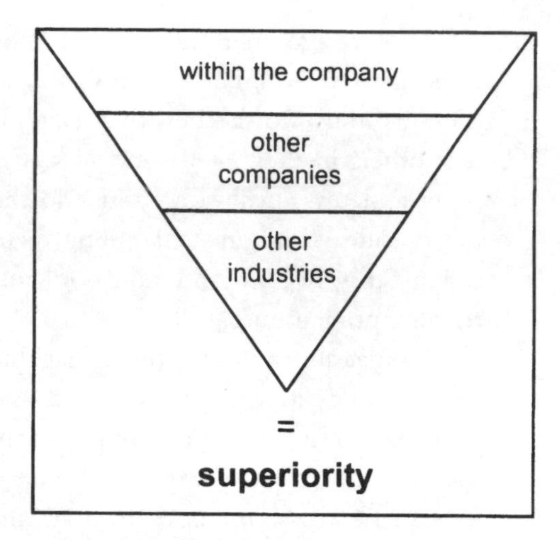

Fig. 4.1 Benchmarking: The pyramid of success.

It is structured

Some managers will instinctively compare their departments or companies with others. Comparison may also happen inadvertently when, for example, trade journals publish league tables or statistics. This however is not benchmarking. Benchmarking is a management technique aimed at achieving superiority. As such it must be a formal and structured approach that is planned, implemented and monitored.

It is ongoing

The business environment within which a company operates may change quickly. New products, processes and techniques are constantly emerging and in order to maintain a position of superiority companies must respond to these changes. For this reason benchmarking, in order to be effective in reaching its aim of superiority, must be a continuous process.

The comparison is with best practice

If a company simply compares itself with others it will not necessarily improve. A critical factor in benchmarking therefore is the search for best practice with which to compare. It is only by benchmarking against best practice that superiority can be achieved.

Its aim is organisational improvement through the establishment of achievable goals

Benchmarking against best practice illustrates to a company the difference between its own practice and best practice. As such it allows the gap in performance to be gauged and for targets to be set aimed at closing the gap and eventually achieving superiority.

Taking these essential ingredients into account, benchmarking can therefore be defined as:

> A process of continuous improvement based on the comparison of an organisation's processes or products with those identified as best practice. The best practice comparison is used as a means of establishing achievable goals aimed at obtaining organisational superiority.

Historical development

The meanings of the words 'benchmark' or 'benchmarking' are known to most people in the construction industry as a level against which other heights are measured. However in management the word has been used in a broader context of setting a standard against which to compare for a considerable period of time. Even the work of Taylor who, as far back as the 1800s, encouraged the comparison of work processes, has been compared with benchmarking[4]. The term benchmarking is also used extensively in the computer industry to illustrate a level of performance of software or hardware[5].

The word benchmarking as defined above and used in the context of this book is a new management technique 'invented' by the Xerox Corporation in 1979[6]. Xerox were aware that photocopiers were being sold in Japan for the same price that they could be made in America and they wanted to find out why. They therefore bought the rival copiers, stripped them down and analysed their component parts. This process proved a successful means of improvement and Xerox therefore extended the use of the benchmarking technique to all business units and cost centres of the company.

The non-manufacturing departments within Xerox initially found difficulty in applying the benchmarking technique, since as they only dealt in business processes they had no product to strip down and compare. However they eventually recognised that processes were the means by which the final product was delivered and that these processes could equally be compared with the external environment as a means of bringing about improvement.

Despite its success at Xerox, benchmarking remained outside of the public domain for some years. Two events helped to change this. First was the book written by Robert Camp in 1989. Camp had worked with the benchmarking initiative at Xerox for seven years and in 1989 formalised his ideas in *Benchmarking: The Search for Industry Best Practice that Leads to Superior Performance*. Second was in 1992 when the Malcolm Baldrige National Quality Award, which is a prestigious quality award given to American companies, introduced a category of benchmarking and competitive comparisons as a criterion of the award. These two events therefore brought the subject of benchmarking into the public domain in the USA[7].

The increase in importance of benchmarking in the USA was followed by an increased interest in the UK and in Europe. In the UK for example the Department of Trade and Industry developed a business-to-business exchange programme offering visits to UK exemplars of best practice in manufacturing and service industries. For a nominal fee companies may visit the host organisation with the objective of transferring best practice to their own organisation[8]. The European Foundation for Quality Management now also recognises the importance of benchmarking[5].

As is often the case with the introduction of new management techniques the construction industry lags behind that of manufacturing. However benchmarking is now beginning to be researched and used within the construction industry as well. The best examples to date are the Construct IT Study[9] and the work by the Building Research Establishment[10]. The former involved benchmarking 11 leading construction companies' use of IT in site processes against a company identified as best-in-class (best practice). The BRE work is concentrating on producing a benchmarking methodology for the construction industry.

Types of benchmarking

Earlier in this chapter (Figure 4.1) the benchmarking pyramid of success outlined that an organisation in the search for best practice against which to compare could look to internal sections or divisions, other companies, or other industries. Naturally these three types of comparison would involve different procedures and would, in addition, offer different benefits and disadvantages. For this reason they are generally classified as three distinct types of benchmarking, each of which is examined separately below. Before this examination however, another concept, central to the idea of benchmarking, must first be discussed. This is the concept of the business process.

Any organisation is broken down into a series of functions. In business terms function refers to the performance of a particular section of the organisation such as marketing, estimating or buying. All functions have an output or deliverable. In the case of the estimating function for example, the output may be the total number of submitted bids. A business process, on the other hand, refers to the action that takes place within the function. In the case of estimating, the business processes may therefore be the decision to tender, the obtaining of subcontractor quotes or the final submission of bid. Within these processes there will also be subprocesses. In the case of final submission of bid, these subprocesses might include the checking of subcontractor bids, the calculation of attendance on subcontractors, the addition of contingency, the addition of overheads and profit and submission to the client. A process differs from a function in that it is a state of being in progress or that which converts input to output. The sum total of the outputs of all the processes is the product or deliverable of the function. The sum total of all products delivered by all functions is the final product, which in the case of construction is the completed building.

In the example above, the output of the estimating function can be viewed quantitatively in the amount of projects won as a percentage of those bid for. This could then be compared with other contractors to see if there was a performance gap. This type of quantitative analysis in benchmarking is called a metric. The problem is that even when a metric indicates that a performance gap exists, it gives no indication of why. If on the other hand the business processes were analysed, the reasons for the performance gap would be clear from the outset. Most benchmarking texts therefore recommend that processes be examined in preference to metrics.

This however leads to a tautological problem of how a company can know there is a performance gap unless they examine the metrics first. This is where benchmarking requires a shift in the usual mode of thinking. For reasons that are beyond the scope of this text, metrics and concentration on them as a means of improvement, can tend to misdirect effort. If on the other hand the business processes and subprocesses are viewed as pieces of a jigsaw, with the picture being the function products and ultimately the final product, then it can be seen that an improvement in all the processes, or at least those most critical to the success of the organisation, will lead to an improvement in the final product. Processes give products. It can therefore be assumed that best possible processes will lead to best possible final products.

In manufacturing some benchmarking may still be based on the comparison of final products and their components. However in construction management and in this text, the word benchmarking is assumed to relate to the process only. In this context there are three types of benchmarking.

Internal benchmarking

This is the comparison of different processes within the same organisation.

As outlined earlier, an essential component of benchmarking is the search for best practice in the external environment. If this is so, then why would an organisation carry out internal benchmarking? The answer to this is fourfold. First, it is possible that within the same organisation business processes will vary. This may be for reasons of location or may be historical, the company having been subject to takeover bids or mergers. Internal benchmarking gives the organisation an understanding of its own performance level. It allows best practice that exists within the organisation to be identified and installed company wide. The second reason is that internal benchmarking provides the data that will be required at the 'external' benchmarking stage. Third, internal benchmarking, by encouraging information exchange and a new way of thinking, ensures that the process of benchmarking is understood by those who will be involved in later 'external' benchmarking exercises. Finally, benchmarking is based on the examination of the business process, as such comparison with other divisions of the same company, although internal to the organisation, may provide a comparison that is external to the process under consideration.

An example of internal benchmarking would be a construction company comprising a major works division, a housing division and a refurbishment division, comparing the way the three divisions deal with the hiring of plant.

Competitive benchmarking

This is a comparison between the processes of companies operating within the same industry. The big advantage with this type of benchmarking is applicability. It is highly relevant to compare the marketing operations of two companies offering the same product and working within the same client base. The problem however with competitive benchmarking is that because we are dealing in process, the best practice of a competitor is not necessarily good enough. A particular construction company may, for example, have an excellent reputation for design and build projects. There is however no direct follow-on from this that their estimating processes are any better than others. As a result benchmarking such processes will not create superiority. In order to identify best practice in the business process, it is sometimes necessary to go beyond the sphere of one's own industry.

Generic benchmarking

This compares the business processes of organisations regardless of the industry they belong to. Some business processes are common to all industries: purchasing and recruitment are two examples. The advantages of generic benchmarking are that it breaks down the barriers to thinking and offers a great opportunity for innovation. It also broadens the knowledge base and offers creative and stimulating ideas. The disadvantages are that it can be difficult, time-consuming and expensive.

Before moving on to examine how the technique of benchmarking operates, two other items need to be considered. The first is whether internal benchmarking is prerequisite to competitive benchmarking and whether that in turn is prerequisite to generic benchmarking.

The basic answer to this is no. The benchmarking pyramid of success shown earlier is an ideal situation; all organisations should fully understand both their own processes and those of their industry before they begin to examine those of other industries. However it is possible, although not recommended, that a company could carry out a generic benchmarking exercise without having carried out either internal or competitive benchmarking. In addition, a company may carry out only internal benchmarking without recourse to the other two types.

The second item which needs to be mentioned is that not all current benchmarking texts use the same terminology and this may cause a certain amount of confusion. Table 4.1 summarises the different terminology used in some of the major benchmarking texts.

The fact that different terminology is used to describe what is essentially the same does not matter. The important item is that an organisation selects the terminology it is most comfortable with and

Table 4.1 Benchmarking terminology.

Author	Within the organisation	Product-to-product comparison	Different companies in the same industry	Different industries
Camp[6]	Internal	Competitive	Functional	Generic
Spendolini[7]	Internal		Competitive	Functional (Generic)
Karlof & Ostblom[11]	Internal		External	Functional
Blendell *et al*[12]	Internal		Competitor and functional	Generic
Copling[5]	Internal		External or best practice	External or best practice
*Watson[4]		Reverse engineering	Competitive	Process
Peters[13]	Internal	Benchmarking	Benchmarking	Benchmarking
This text	Internal	Not applicable	Competitive	Generic

(* Watson also includes two other categories of strategic benchmarking and global benchmarking.)

uses this consistently. The choice may or may not correspond with the terminology of this text summarised in Table 4.1

Figure 4.2 below shows that as the type of benchmarking moves from internal to generic, the level of difficulty, the time taken and cost incurred increase along with the creativity and the opportunity for improvement. Conversely, when moving from the generic to internal benchmarking, cost, time and difficulty decrease as do relevance, ease of data collection, applicability and transferability of results.

The process of benchmarking

Although the methods suggested by the major texts for implementing benchmarking studies are numerous, all the methods contain the same essential ingredients. This book suggests a nine-step

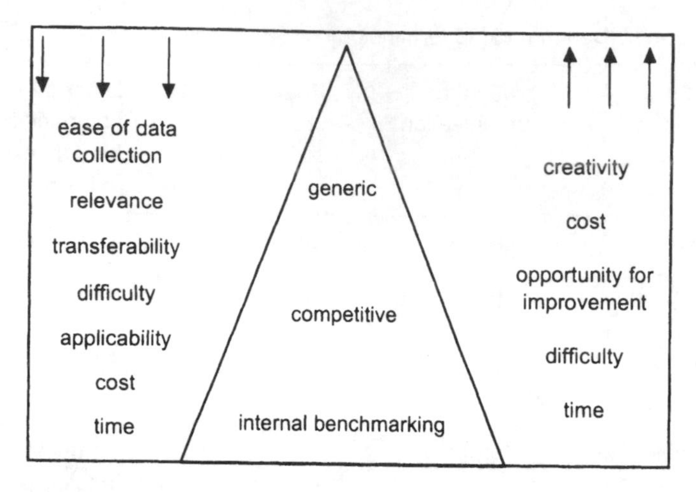

Fig. 4.2 Types of benchmarking.

approach to benchmarking, as shown in figure 4.3, and each step of this approach is examined separately.

Step 1: decide to benchmark

The free market is dependent on the customer being able to choose between alternative suppliers[9]. This means that when an organisation supplies goods and services that are not wanted, or not of an acceptable standard, customers will not buy them and the company will go out of business. This same freedom of choice does not exist within an organisation. The internal nature of the environment means that where goods, or more likely services, are offered by one department to another, there is no alternative available if that service proves to be unsatisfactory. What this in effect means is that although a company may appear profitable and geared to user needs, there may in fact be scope for improvement within the internal processes of the organisation. Improving these internal processes will increase efficiency and ultimately improve the standard of the final product. Benchmarking therefore provides a 'safety net' for those processes which are not exposed to market forces. As illustrated above, efficiency decreases with distance from the final product[12], and this is the major reason for carrying out benchmarking. However there are other reasons, some of which are outlined below.

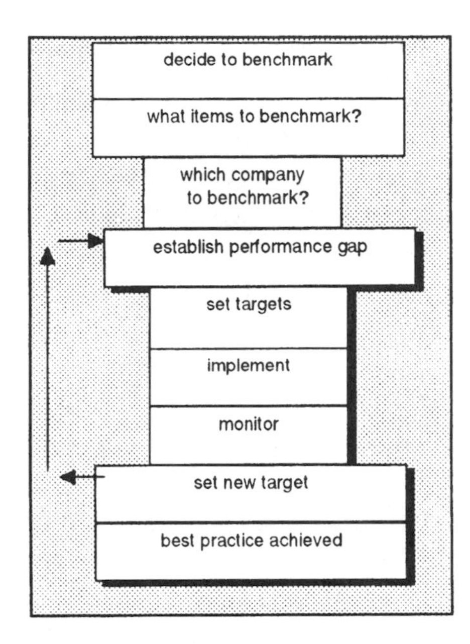

Fig. 4.3 Nine-step benchmarking approach.

Speed of change

Today's business environment changes more rapidly than ever before. In the construction industry new procurement methods, new products and new clients are constantly coming to the fore. All of these exist in the external environment. A company that does not look to the outside for these will be overtaken by the pace of change, stagnate and eventually go out of business. One of the essential ingredients of benchmarking is comparison with the external environment and as such the technique forces a company to constantly appraise the changing situation of the industry.

Free exchange of information

In Britain business culture is such that information is not exchanged freely and there tends to be an underlying assumption that most information is confidential. This is totally different from the business culture that exists, for example, in the USA, where information is freely exchanged. The onset of the 'information age' will undoubtedly require a shift in the way information is regarded in Britain and the use of benchmarking encourages a freer exchange of information.

External focus and meeting of customer objectives

In order to stay in business the main objective of any company must be to meet customer requirements. However, as with change, customer needs and markets are external to the company. An internally focused company cannot understand its customer needs and cannot therefore meet the demands of its markets. Benchmarking encourages the external focus required to meet customer objectives.

The recognised importance of the process

After World War II the need to provide high quantities of consumer goods was the main objective of manufacturing companies. Even in the construction industry emphasis was on the provision of large amounts of accommodation, both for residence and industry. This situation has changed gradually over the last 30 years. Japanese industry was quick to recognise this change and began producing goods of a much higher quality than its competitors, resulting in a huge expansion of Japan's manufacturing base. Examination of Japanese methods of manufacture showed a much greater emphasis on the processes, which ultimately deliver the products. British industry is now imitating this with initiatives like TQM, reengineering and benchmarking, all of which recognise the importance of improving the final product through examination of the process.

Highlighting the performance gap illustrates the need for change

Traditional methods of business evaluation tend to compare the performance of individual departments within the same company. Alternatively performance evaluation is based on a comparison with previous years or against a forecast of predicted performance. This type of evaluation is internally focused. Since the measure being used is either the best the company has at present, or the best they have achieved in the past, such an approach may encourage companies to fall short of the performance they are in fact capable of achieving. The true measure of a company's performance can really only be gauged by comparison with its competitors. Benchmarking, by operating in the external environment, illustrates the true performance gap. In doing so it highlights the need for, and the motivation for, change.

Best practice reveals how change can be achieved

In benchmarking the term 'metric' is used to describe the quantifiable output of a process. For example, two construction companies will have different processes for the calculation of a tender sum based on a bill of quantities. One of the metrics that results from these processes may therefore be the 'hit rate' of successful tenders. If the two metrics were compared, a performance gap could be highlighted; but what use is this information in isolation? Benchmarking discourages the use of metrics as a means of comparison since they only show the performance gap. If however one of the contractors' processes had been identified as best practice, then the metric largely becomes irrelevant. Processes deliver output, in this case the tender sum. Selection of the best possible process will therefore automatically lead to the best output. It is not therefore necessary to concentrate on the metric. A further advantage of concentrating on the process is that unlike the metric, which only indicates the performance gap, it shows why that gap exists. As a direct consequence it shows how the gap can be closed. Benchmarking the process therefore not only highlights the problem but also provides the solution.

Best practice indicates what is achievable

An obvious question to ask in relation to benchmarking is why stop at best practice? In the same way that benchmarking against a company's best internal practices may be encouraging it to fall short of its potential, then so may be the benchmarking of best practice. There is no direct follow-on that external best practice is the best that could possibly be achieved. In addition, in following the best practice of another company, there is an argument that a company is lagging behind instead of leading from the front. The answer to these arguments is twofold. First, benchmarking stresses process and within an organisation there may be hundreds of processes and subprocesses. A company is therefore following only in regard to processes, the combination of which will give it business superiority. Second, benchmarking best practice shows what is achievable. If a company were to aim higher than best practice there can be no guarantee that it could be achieved.

Benchmarking provides an environment for change

As outlined earlier, benchmarking requires a shift in business culture. In order to be effectively implemented it requires a change in the way information, internal business processes and competitors are viewed. The problem with this is that with most organisations there is an intrinsic reluctance to change. Advocates of benchmarking argue that the actual process of benchmarking itself can help to overcome this resistance. Because benchmarking is a creative process that investigates how other organisations carry out their business processes, it acts as a catalyst to change. Because benchmarking clearly shows when a performance gap exists, it also helps to motivate employees towards making the change necessary to close the performance gap.

Benchmarking identifies the technological breakthrough of other industries

This is often given as one of the advantages of benchmarking and the example usually cited is that of barcodes. Although this was first a technological breakthrough of the grocery industry, barcodes are now also used in libraries, hospitals, security and identification systems.

Benchmarking allows individuals to broaden their own background and experience

People operating in a particular industry tend to adopt the business culture of that industry. Although this is necessary in order to work effectively within that industry, the business culture may also present restrictions and stifle change. Working with other organisations and industries illustrates that there is more than one way of carrying out any task and that existing methods can almost always be improved.

Benchmarking focuses on the objectives

Benchmarking highlights the performance gap and sets targets aimed at closing that gap. These targets then become a focus or objective for those involved in the benchmarking process. When all staff are focused on the same objectives, goals are more likely to be achieved.

The industry best is the most credible goal

One of the problems with implementing successful manufacturing techniques in the construction industry is the claim that the construction industry is different and not conducive to the application of manufacturing techniques, however successful they may have been. A business process however exists regardless of industry, and focusing on the best practice of such processes provides a credible goal which can be recognised as achievable. Even for the construction industry!

Step 2: what to benchmark?

This is probably the most difficult part of benchmarking. It is maybe for this reason that the main benchmarking texts do not concur on what in fact should be benchmarked. This text has stressed that one of the fundamentals of benchmarking is process; however not all of the texts agree with this. Peters[13] for example defined three levels of benchmarking: strategic, operational and statistical. Strategic benchmarking deals with benchmarking culture, people, skills and strategy. (Pastore also agreed that benchmarking could take place on strategy[14].) Operational benchmarking deals with benchmarking methods, procedures and the business process. (This is also called process benchmarking.) Finally statistical benchmarking is the numerical or statistical comparisons of company performance.

This idea that more than the process can be benchmarked only succeeds in complicating what is essentially a simple issue. 'Best practice culture' cannot be benchmarked. Even if it could there would be little purpose in doing so, since culture cannot be changed easily, if at all. Culture, including the subculture which may exist within an organisation, is an intangible asset[15]. In addition there is a close relationship to the type of product a company makes and its corporate culture[16]. As such there is little purpose in benchmarking culture when the products offered are likely to remain different. Making a separate category of benchmarking to deal with purely statistical or numerical comparisons also complicates the issue. As explained earlier, the numerical measure of performance or metric is an output of the process. Comparison of these metrics achieves little other than stating that a performance gap exists. As benchmarking is defined as a process of improvement, such statistical comparison alone could not be called benchmarking.

There is of course a strong possibility that the technique of benchmarking will develop and it may, at some future date, be feasible that some aspects of corporate strategy or even of people may be benchmarked. However in the context of this text and, the authors believe, in the current and accepted context of benchmarking, the technique embraces only the comparison of products or processes. In the narrower context of construction management the term benchmarking includes the comparison of business processes only, the final product (the building) being too diverse and complex to facilitate meaningful comparison.

How then is a business process defined?

Any business activity can be seen in three stages. There is an input, a process and an output. The combination of these outputs leads to the final product. For the purpose of organisation and administration certain outputs are achieved by grouping them into a business function. Figures 4.4 and 4.5 show how this idea operates in the case of a building contractor. Figure 4.4 shows how the company organises itself into functions, within which processes take place. Figure 4.5 shows in detail the processes and subprocesses that take place within the 'subcontractor management function'.

In figure 4.5, project management is the business function and subcontract management is a process within that. Within this process there are, among others, the subprocesses of subcontract tendering and sitework. Within the process of tendering there are further subprocesses of keeping an up-to-date tender list and obtaining tenders. These subprocesses take a certain input in terms of resources and in addition produce an output. In this case the output is the completed tender list and the tenders received from the subcontractors. The sum of all these outputs is the process output, the sum of which is the function output. The sum of all the function outputs is the final product: that is, the completed building. This is

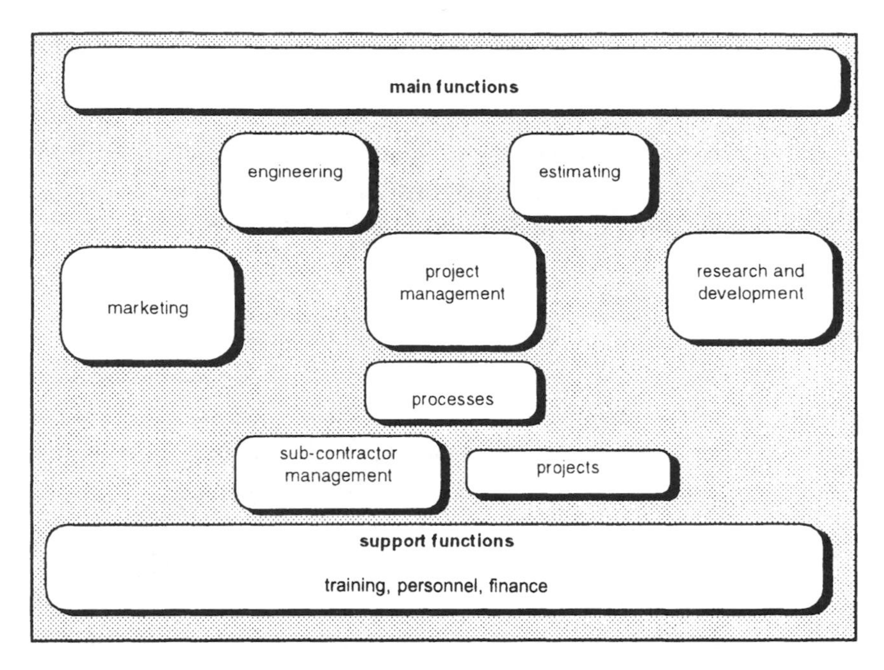

Fig. 4.4 The functions of a contracting organisation.

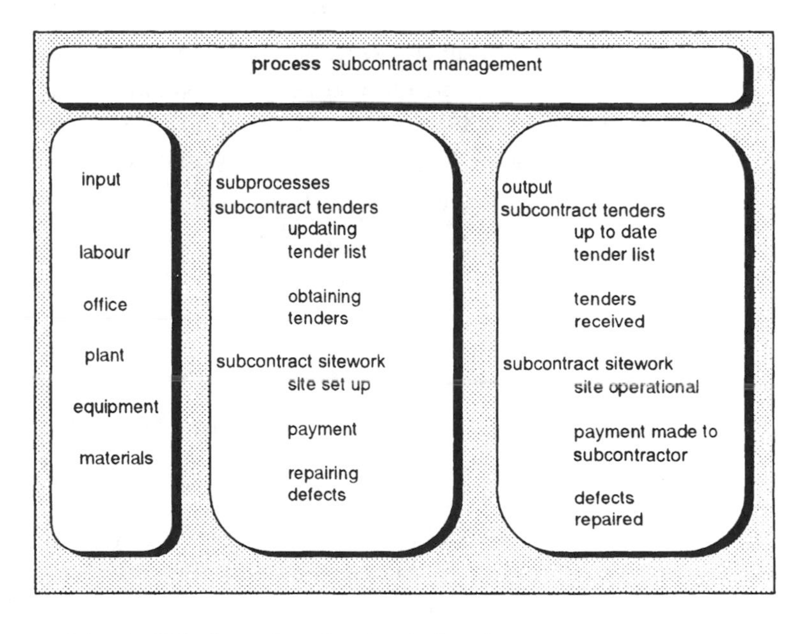

Fig. 4.5 Processes contained within the subcontract main function.

the basis of benchmarking: to improve the process so that the final product is improved.

The problem that arises however is that as an organisation may have hundreds or even thousands of subprocesses, how to decide which ones should be selected for benchmarking? There are various ways this can be done and these are outlined below.

Identify the product of the business function

As explained earlier the organisational or administrative divisions within an organisation are its business functions. These can be a good starting-point in deciding what to benchmark, since business functions tend to be much more visible than business processes. It may for example be known that contractor X has excellent project management. Another contractor, aware his own project management is poor, may use this business function as a starting-point for his benchmarking exercise. Investigation of the process within that function may lead him to conclude that his real problem is his subcontractor management. He may therefore choose to do a detailed analysis of his subcontractor-management process and compare it with the more successful contractor. Identifying the superior product function therefore leads to the superior process which produces it.

The problem with this approach is there is no definite boundary between function and process. Subcontract management had been given the title of process in the example above but it could equally be described as a function. Although the division of process and function is often obvious, there is also a grey area where no definite distinction exists. In such circumstances the only answer is experience of the organisation, industry and management.

Critical success factors

Within any organisation there are critical success factors. These are those factors which, if not operating effectively, will damage the operation of the organisation. In the case of the housebuilder, marketing may be a critical success factor, whereas in the civil engineering division of a major contracting organisation, a critical success factor may be completion of projects on time. Identification of these factors can be a good starting-point for benchmarking. In addition to being critical success factors, items selected for benchmarking must be those in which an improvement will make a significant difference to the overall performance of the company. For example a contractor may know that turnover is a critical success factor but if the company is already operating at capacity an improvement in that area may not substantially benefit the organisation. Critical success factors can be likened to the Pareto rule which states that 80% of the result is based on 20% of the activities. For benchmarking purposes the critical success factors are those 20% activities. It is surprising how few construction managers do actually understand the critical success factors of their own organisation. As a result when organisations get into difficulty managers embark on cost-cutting exercises, as this is the only way they can see of improving performance. They believe that reducing cost while maintaining output automatically adds value. In reality such exercises achieve little other than deflating morale and reducing the quality of service offered.

The Xerox questions

To assist in the identification of what should be benchmarked Camp[6] recommended a series of questions, most of which are reproduced below.

What factors are most critical to your business success?

In a contracting organisation the following may be critical success factors:

- Turnover
- Tender hit rate
- Ratio of civil engineering to building projects
- Number of houses sold

What areas are causing the most trouble?

In this case the answers may be:

- Management of subcontractors
- Meeting completion dates

What products or services are provided?

Typically this might be

- ❑ Houses
- ❑ Maintenance
- ❑ Small building projects

What factors are responsible for customer satisfaction?

If clients of the construction industry were surveyed, the following might relate to their satisfaction:

- ❑ Completion on time
- ❑ Minimum cost
- ❑ Guaranteed cost
- ❑ Quality
- ❑ Health and safety
- ❑ Environmentally friendly development
- ❑ Architectural expression
- ❑ Life cycle cost
- ❑ Durability

What are the competitive pressures?

These may be regional, national or international. They may be coming from smaller builders forcing their way into new markets or large contractors looking for smaller projects in a time of recession.

What are the major costs/Which functions represent the highest percentage of cost?

This may be the head office, permanent staff, manufacturing subsidiaries or marketing.

What functions have the greatest need for improvement?

This could be staff training or the ordering of materials.

Which functions have the greatest effect or potential for differentiating the organisation from competitors in the market place?

This is often the most difficult question to answer in that it tries to identify what gives one company the edge over another.

Ask the customer

Another way of deciding what to benchmark is to ask the customer. In the example used earlier a company may see project management as an area of potential improvement. But who is the customer for this function? Is it the client, the site agent or the subcontractors? The answer is most likely the latter two. Speaking with these will highlight where the problems of management are and these may become the areas targeted for benchmarking. 'Customer' does not necessarily refer to the final user. In the case of benchmarking, customer refers to the customer of the process; that is, the person who is next in the chain and who receives the process output.

It can be seen that there is no definite answer to the question of what processes to benchmark. Any of the above methods or any combination of them could be used to determine the answer. Benchmarking, like many of the concepts in this book, is a soft system. Answers will therefore be highly company-specific and dependent on the organisation's culture. The only real answers come from a thorough understanding of the organisation.

A further question which may be asked when deciding what to benchmark is what is the level of detail that is required. As shown above, a business process may be broken into many subprocesses and these subprocesses may be broken down still further. The level of detail selected is entirely up to the individual organisation and will vary according to the size of the organisation and the amount of resources they wish to dedicate to the benchmarking activity.

Step 3: which companies to benchmark?

The answer to this question is simple yet at the same time very difficult. The simple part is that the comparison or benchmark company is one whose processes are best practice. The difficulty is how to find out who these companies are. In some functions, such as marketing, finding out who is the industry best can be quite straightforward. In the case of housebuilding particularly, information on sales, advertising and brochures can be collected easily. In addition sites can be visited and showhomes and site marketing examined. If resources are available house purchasers can be surveyed and a marketing consultant employed to analyse the results. Once all available information is collected, a judgement of best practice can be made.

In the case of a civil engineering division of a major contractor the process of identifying best practice becomes much more difficult. The product, not being something that the general public purchase, is further removed from the market place. The number of finished products is limited, there are no brochures and little data about projects in the public domain. How then can the information regarding best practice be obtained?

The only way to establish leading companies in a specific function or process is through a collection of data. Most of the benchmarking texts give a list of sources of information to locate best practice companies and the major sources are reproduced below.

- Special awards and citations
- Media attention
- Professional associations
- Independent reports
- Word of mouth
- Consultants
- Benchmarking networks
- Internal information
- Company sources
- Library
- Experts and studies
- Questionnaire by mail and telephone
- Direct site visits
- Focus groups
- Special interest groups
- Employees, customers and suppliers
- Foreign data sources
- Academic institutions
- Investment analysis

- The internet
- On-line data base
- Journals.

Step 4: establishing the performance gap

Once best practice has been identified the company who owns it needs to be approached to be a benchmarking partner. To most construction management professionals this presents the biggest problem of benchmarking, in that there is a reluctance to approach other companies or competitors. Likewise there is also a reluctance on the part of the company approached to give information which will be used as a basis for improving their competitors. As outlined by Spendolini[7], most people employed in private enterprise see it as their role to beat the competition, not train them.

This attitude is understandable; after all it is based on years of industrial practice. However in reading all the sections of this book one item should by now be clear. That is that all of the concepts presented, be they TQM or value management, require some sort of cultural shift in the way the construction industry operates. Markets in all industries have changed beyond recognition since the early 1980s. Markets are now global and customer-focused. The construction industry has been slow to recognise this but external competition means it cannot ignore it any longer. If the industry is to survive, it must change. A part of this change will undoubtedly demand some degree of co-operation between companies.

There are no definite rules on how a company should be approached with a request to become a benchmarking partner; this is a question for the individual company concerned. However, the critical factor is that it is a partnership. Both parties should expect to gain something from the experience, even if it is how to conduct a benchmarking exercise.

Having identified best practice and entered into a benchmarking partnering agreement the next step is to fully document the process that is under study. There are many methods of doing this, such as flow charts or process diagrams. Space does not allow a detailed examination of them here, but all good management texts will usually explain these techniques thoroughly.

In earlier stages of this chapter the use of the metric or quantification of the process output was discouraged. However now that a thorough examination of the process is documented the metric can be used. In some instances use of the metric can be highly relevant; examples are tender hit rates, projects completed on time or claims made by subcontractors. In some instances however it may be sufficient to simply document the process and compare the processes under study. The movement of a tender within an organisation is a good example. Where a metric is used it can take several forms: ratio, cost per contract, cost as a percentage of turnover. Whichever method is used the company will form one of the three judgements about its own performance of a particular process in relation to that of the benchmarked company:

Parity: that both companies have equal performance.
Positive performance gap: that the company under study is in fact performing better than best practice.
Negative performance gap: that the company under study is performing below the level of best practice.

Step 5: set targets

The most likely outcome will be that company performance is below that of the benchmarked company and that there is a negative performance gap. In such instances the company will need to devise how the gap will be closed. Unless the gap is small and can be closed in one company incentive, it will be necessary to set a target aimed at achieving the best practice in a series of stages.

It is not the intention of this text to give a detailed account of the procedures for goal-setting within the organisation, since these can be found in any good management text. Suffice to say the target must be planned, realistic and achievable. It must be communicated to staff and require their participation.

Steps 6, 7 and 8: implement-monitor-set new target

The next stages of the benchmarking process require that the plan to achieve the first target be implemented and monitored, followed by the setting of a new target. Once again procedures for doing this are not included here since they can be found in most management texts. In addition most companies will have their own procedure for implementing plans and monitoring progress. Lack of inclusion here is not intended to reduce the significance of these stages of the benchmarking process. They are of course vital to its success. A plan that is made and not implemented is worse than no plan at all.

Step 9: best practice achieved

Best practice achieved means the company has achieved parity with the benchmarked company. However, the benchmarking process does not end here. As outlined earlier, benchmarking is a continuous process. Like TQM it is something a company has, not something a company does once in a while. In order to achieve superiority a company needs best practice in all of its major processes. It can only achieve this by constantly looking for further improvements.

The first section of this chapter has looked at the steps to be taken in a benchmarking study. The second section deals with items relevant to the implementation of the study.

The benchmarking team

Benchmarking, unlike value management, is not necessarily a team activity; it could be carried out by an individual. However, in the interest of sharing the workload and also as an aid to acceptance and communication of the benchmarking, a team is the recommended approach.

There are various formats of a benchmarking team.

Work groups

These are groups that already exist and may comprise members of an existing department or group.

Interdepartmental groups

These are formed by bringing together people with particular expertise in the area required. These teams will generally disband after the benchmarking study is complete.

Ad hoc teams

These are more flexible and made up of any person interested in joining the benchmarking group. Group members may come and go. Benchmarking groups of this type will not work well unless the company is a mature benchmarking company such as the Xerox Corporation.

Benchmarking Code of Conduct

The American Productivity and Quality Centre's International Benchmarking Clearing House and the Strategic Planning Institute Council of Benchmarking have developed a code of practice for benchmarking[4]. It contains nine basic principles as outlined below.

Legality

This principle excludes anything contrary to restraint of trade, such as bid-rigging.

Exchange

Benchmarkers should not ask another company for information that they themselves would be unwilling to share.

Confidentiality

Nothing learned about a benchmarking partner from a benchmarking study should be shared with anyone else.

Use

Information gained from benchmarking studies should not be used for any purposes other than process improvement. Advertising is the most obvious example.

First-party contact

The first contact at the company to be benchmarked should be the person responsible for benchmarking, not the person in charge of the process.

Third-party contact

Company contacts should not be shared with other companies without first gaining permission.

Preparation

Benchmarking partners should not be contacted until all preparatory work is complete.

Completion

Commitment to benchmarking partners should not be made unless it can be followed through to the completion of the study.

Understanding and action

Before embarking on a benchmarking study both companies should understand the process and have made their intentions clear, particularly about the use of information.

Legal considerations

The following are the major legal considerations of benchmarking[17].

- Chapter 1 of Part III of the Rome Treaty, particularly articles 85(Sub 1,2,3) and Regulation 17 (article 4).
  These agreements apply to all EU countries and state that agreements of price-fixing, sharing markers, discrimination against third parties to their competitive disadvantage or imposing territorial restrictions which partition the common market are prohibitive.
  Domestic legislation includes:

- The Restrictive Trade Practices Act 1976
- The Fair Trading Act 1973
- The Competition Act 1980.

Benchmarking: the major issues

It is not the intention of this book to 'sell' the technique of benchmarking. Rather it is to give an objective appraisal of how the technique can operate in the construction industry. In that vein the following is a summary of current thinking which those concerned with more academic aspects of benchmarking may find interesting.

The relationship between benchmarking and TQM

Most benchmarking texts agree that benchmarking works best in companies with a culture of total quality management. Jackson *et al.*[18] argued that without a total quality control programme, benchmarking is a waste of time and money. Whether or not this is true, particularly in the construction industry, is really not yet known.

The place of innovation in benchmarking

One of the criticisms of benchmarking is that because it follows as opposed to leads, it is not an innovative technique. Watson[4] however saw benchmarking as only part of the equation and that superiority lies in quality beyond the competition, technology before the competition and cost below the competition. The authors agree with this. Benchmarking is not, or was not intended as, a tool of innovation, but of process improvement. Blendell *et al.*[12] however, in contrast to this view, saw that the technique could be innovative and that a fourth type of benchmarking, customer benchmarking, goes beyond what the customer expects or requires. In this way it is an innovative process and not only one of process improvement.

The problem of using metrics

A paper in *World Wastes*[19] used the metrics of waste services at several major American cities. This is a classic example of why metrics should be avoided when benchmarking. The paper described how benchmarking was used 'to measure one service delivery system against another', and compared the number of refuse collections along with the costs of waste services. However as this process neither selected best practice nor aimed at process improvement, it could not be described as benchmarking. It was a simple comparison of quantifiables.

Although theoretically there is clearly logic in delaying the use of metrics until the business process is documented, the reality is much more difficult. Managers will want to see at least an estimate of the size of the performance gap before they allocate resources aimed at closing it. Arguments about the need for a cultural shift will have little effect on the manager with limited resources. The answer here is that metrics can be used earlier in the benchmarking process if there is no alternative. However they should always be used with caution and, on documentation of the process, be recalculated.

Business co-operation

For many managers this is the most difficult aspect of benchmarking. There are two basic arguments in favour of business co-operation. First, the essence of benchmarking in construction lies in the process, which is not as sensitive as the product. Second, and strongly in keeping with the theme of this book, there is a desperate need for a cultural shift in the construction industry. Benchmarking, according to Watson[4], is about a coherent *national challenge* for continuous improvement. Many companies already recognise that future success depends on the global market place and the development of national productivity. Without co-operation there can be no benchmarking. Any company who cannot tolerate this concept will not therefore be able to carry out the benchmarking process.

How far to go?

In the search for best practice how far should a company go? Should, as recommended by Grinyer and Goldsmith[20], the search be a global one? The answer to this must realistically be no. It is not feasible for a medium-sized building contractor to search the world as a means of finding improved processes. Very few organisations are capable of such an extensive and exhaustive search. The benchmarking partner must clearly be a comparable size and status company that offers processes with a high degree of applicability and relevance.

Japanese kaizen

Benchmarking is, in some texts and articles, compared to Japanese *kaizen*[21]. This however is not a correct comparison since *kaizen* is a much broader concept that does not rely solely on a system of external comparison. *Kaizen* is examined in greater detail in the chapter on total quality management.

Disadvantages of benchmarking

Lenkus[22] looked at some of the disadvantages of benchmarking: namely, that it is complex, cumbersome, time-consuming and expensive. He also found problems with securing benchmarking partners. Although he could see the advantages of benchmarking metrics he saw that benchmarking processes was difficult. He also believes that managers invariably want to see metrics before they will commit resources to process improvement. Other disadvantages included the length of time taken and the fact that technological change may move faster than the time it takes to carry out a benchmarking study.

Current research

Benchmarking, particularly in construction, is relatively new. As such there is still much research to be carried out on the technique. The following deals with selected findings of work that has been undertaken.

Benchmarking has been carried out on the pre-planning phase of construction projects[23]. This work looked at four major subprocesses in the pre-planning process of 62 construction projects. These were to organise, select project alternatives, develop a project definition package and decide whether to proceed with the project. The work collected data on these processes and analysed the data. There are several contentious issues with this work. First, the process and subprocesses examined were based on a theoretical model against which the 62 projects were compared. However as the model is purely theoretical no judgement about project performance can be made in relation to it, since if the model were incorrect then so are all the resulting judgements. The second question which arises from the research is whether a construction project and its processes are in fact capable of being benchmarked. The number of design and constructing firms involved make it an incredibly complex process[24]. A construction management team is in effect a temporary organisation with one-off processes established on an ad hoc basis. Given a limited design and construction span and the fact that many organisations input into the processes, how could best practice ever be established? Even if a project's processes were identified as best practice, the introduction of a new client, consultant or contractor would surely make repetition of them impossible. This raises a very serious question for the future of research into benchmarking in construction: namely, can the project be benchmarked or can benchmarking only apply to the individual organisations that exist within the project?

Another interesting aspect of benchmarking research is the

relationship of the technique to the culture within which it exists. Watson[4] recognised that given that the future of benchmarking lies in its global application, benchmarking cannot be viewed outside the contexts of international trade and culture. He sees a need in the future of benchmarking to bridge those cultural gaps. However, national differences are not the only culture that may affect benchmarking. Company or corporate culture may also be related to it. Companies have three types of performance behaviour [14] which are basic, innovative and competitive. In the innovative sphere a company takes a risk and hopes the customer will like it. The Sony Walkman is a good example of this. With competitive behaviour, competition is based on direct comparison of product features among product alternatives. Unlike in innovative behaviour, the requirements of the customer are not guessed at but are already known. Basic behaviour satisfies the customer's lowest level of expectation and provides only the product features that customers assume will be provided. There can be no question that these corporate cultures exist in the manufacturing field, but do they also exist in construction? If not, is there any value in a construction company benchmarking outside the industry? Do the barriers provided by corporate culture simply become too great?

Spendolini[7] produced an excellent analogy of how cultural differences have an effect on benchmarking. It was what he called thinking out of the box. The further out of the box the company is prepared to go, the greater the cultural differences that are encountered. The rewards however may be greater. This is an interesting idea; however there is no real evidence that it is true, or even possible to achieve.

The work of the Construct IT Centre was mentioned earlier[9]. In this work, 11 construction companies took part in the benchmarking of IT use in construction site processes. The companies nominated their best project from the point of view of IT use. This work used an engineering company as the benchmark, as its business processes were similar to that of the construction industry. The report made several recommendations for effective management of the site process. These included a strategy which elevates the importance of IT, a re-examination of business processes by construction companies, greater investment in IT and greater education of the workforce in the use of IT.

Case studies

In his book Glen Peters[13] examined the benchmarking of customer service. This has traditionally been an area of weakness in the construction industry and as an example the authors carried out a simple exercise of benchmarking: the approach of an average housebuilder to customer service, compared with the approach taken by a leading car manufacturer. Car manufacturers were chosen because the marketing of cars, the product identification and also the customer service provided are visibly superior to anything offered by the housebuilder, yet both are dealing with expensive commodities that are bought fairly infrequently. As explained earlier, since benchmarking concentrates on a process which cuts across industry barriers, it is valid to benchmark the housebuilders' customer service against that of the car manufacturer. The survey questions are taken from the benchmarking example survey contained in Peters' book[13] and are used with his permission. The results are shown in Table 4.2.

Obviously no definite conclusions can be drawn from such a small survey but, as expected, the car manufacturer was more customer-focused than the housebuilder. Not only were they more aware of the costs of gaining and losing a customer but they also had a distinct complaints taskforce that recorded every complaint made. In addition, and unlike the homebuilder, all of their front-line staff had received customer-care training. An interesting aside to this is that a quantity surveying practice was also asked to complete the survey. They returned it uncompleted, claiming they could not understand the questions!

One reason for undertaking our benchmarking study of customer service in the house building sector with the car manufacturing sector was that, at that point in time (1996), there were few examples of benchmarking in the construction industry (with the exception of the IT benchmarking work of Construct IT[9]). However, as McCabe discusses in his seminal work *Benchmarking in Construction*[25], the work of Latham[26], Egan[27] and the Construction Best Practice Programme[28] has given impetus to the uptake of benchmarking in the UK construction industry. McCabe's recent publication contains eight case study examples of benchmarking in the construction industry which is testament to the increased use of benchmarking in the UK since our first exploratory studies in 1996. Our observations

Table 4.2 Benchmarking customer service in the construction and automotive industries.

Question	Car company	Housebuilder
How do you identify who your customers are?	Database Focus groups Quantitative written research Exit interviews	Database Current active accounts Planning leads
How do you gather information about potential customers?	Direct marketing Inserts Through dealers	Written Telephone surveys
How often do you re-examine the information you have about potential customers?	Annually	6–12 months
What percentage of your customers are identified?	Don't know	70%
What percentage of your overall customer/market research budget do you spend on identifying customers?	100%	10%
What percentage of complaints are recorded?	100%	40%
How do you store information about complaints?	Computerised system organisation wide	Manual system on each site
Do you calculate the costs of losing a customer?	Yes	No
Do you calculate the costs of gaining a customer?	Yes	No
Have you identified the areas that produce the most complaints?	Yes	No
What percentage of front-line staff receive training?	100%	80%
What do you do to analyse complaints?	Complaints task force	Front-line employees are asked to recommend change
Do you tell customers what has happened as a result of their complaint?	Yes	Yes and no
Do you offer some sort of compensation to a complaining customer?	Yes	Generally no

are that the culture of benchmarking does not as yet permeate the world at large. In our view the uptake is as yet very patchy.

Conclusion

One of the items that the authors found of greatest interest when writing this chapter was a paper by Yoshimori[29]. In this he looked at how senior managers in different countries regarded the ownership of their companies. In the UK 70.5% of senior managers saw shareholders as their first priority. In Japan only 2.9% took this view. However in reply to whether priority should be given to all stakeholders, which includes employees, 29.5% of British managers responded positively compared to 97.1% of Japanese.

Camp's[6] book which is the first text written on benchmarking outlines the historical development of benchmarking in Xerox. In this he clearly states that the overall aim of the Xerox programme was to achieve leadership through quality and that this was achieved through the three components of benchmarking, employee involvement and the quality process. However most of the texts that followed Camp failed to mention this, concentrating on the technique of benchmarking as a stand-alone technique. Nowhere is this highlighted better than in a paper by Jackson *et al.*[18] which makes an appraisal of the many benchmarking texts currently available. The paper gives a rating to the coverage of certain major benchmarking topics including the total quality management process. However no topic is included on employee involvement as this aspect of benchmarking appears to have diminished in importance. Many of the books on benchmarking hold up the Japanese as a great example of what benchmarking can achieve. Although this may be true, it may equally be the case that Japanese success is partly a function of employee involvement within the organisation. This aspect of benchmarking cannot be entirely overlooked. Even at the 'source' of benchmarking at Xerox, employee involvement was seen one of the essentials of a three-part programme aimed at successful business. This however seems to be forgotten in the development of benchmarking. If it is continually overlooked, it is the authors' view that benchmarking will fail and simply join the ranks of what was termed the flurry of acronymic assaults that have accosted the business world[4].

References

1. Watman, M. (1964) *Encyclopaedia of Athletics*. Robert Hale, London.
2. Temple, C. (1990) *Marathon, Cross Country and Road Running*. Stanley Paul, London.
3. Drucker, P.F. (1994) *The Practice of Management*. Butterworth Heinemann, Oxford.
4. Watson, G.H. (1993) *Strategic Benchmarking. How to Rate your Company against the World's Best*. John Wiley, New York.
5. Copling, S. (1992) *Best Practice Benchmarking. The Management Guide to Successful Implementation*. Industrial Newsletters Limited.
6. Camp, R.C. (1989) *Benchmarking: The Search for Industry Best Practices that Lead to Superior Performance*. ASQC Quality Press, Milwaukee, Wisconsin.
7. Spendolini, M.J. (1992) *The Benchmarking Book*. American Management Association, New York.
8. Department of Trade and Industry. (1996) *Inside UK Enterprise. Managing in the 90's*. Further information can be obtained from Status Meetings Limited, Festival Hall, Petersfield, Hampshire, GU31 4JW. Tel: 01730 235055 Fax: 268865.
9. Construct IT Centre of Excellence (1996) *Benchmarking Best Practice Report. Construction Site Processes*. University of Salford.
10. The Building Research Establishment LINK IDAC (1996) *Project: Benchmarking for Construction*. In Innovative Manufacturing Initiative Report Benchmarking Theme Day, 25 July 1996, London. Published by the Engineering and Physical Sciences Research Council, Swindon.
11. Karlof, B. & Ostblom, S. (1993) *Benchmarking: A Signpost to Excellence in Quality and Productivity*. John Wiley.
12. Blendell, T., Boulter, L. & Kelly, J. (1993) *Benchmarking for Competitive Advantage*. Financial Times Pitman Publishing, London.
13. Peters, G. (1994) *Benchmarking Customer Service*. Financial Times Pitman Publishing, London.

14. Pastore, R. (1995) Benchmarking comes of age. *CIO* **9**, (3) 30–36.
15. Gray, S. (1995) Cultural perspectives on the measurement of corporate success. *European Management Journal*, September, **13** (3) 269–75.
16. Kono, T. (1994) Changing a company's strategy and culture. *Long Range Planning*, October **27** (5), 85–97.
17. Coonen, R. (1995) Benchmarking: A continuous improvement process. *Health and Safety Practitioner*, October, 18–21.
18. Jackson, A.E., Safford, R.R., & Swart, W.W. (1994) Roadmap to current benchmarking literature. *Journal of Management in Engineering*, **10** (6) 60–67.
19. Slovin, J. (1995) Cities turning to benchmarking to stay competitive. *World Wastes*, **38** (10), 12–14.
20. Grinyer, M. & Goldsmith, H. (1995) The role of benchmarking in re-engineering. *Management Services*, October, **30** (10) 18–19.
21. Anon. (1995) Benchmarking Kaizen. *Manufacturing Engineering*, November, **115** (5), 24.
22. Lenkus, D. (1995) Benchmarking support is split. *Business Insurance*, **29** (39), 1, 29.
23. Hamilton, M.R. & Gibson, G.E. (1996) Benchmarking pre-project planning effort. *Journal of Management in Engineering*, March/April, **12** (2), 25–33.
24. Mohamed, S. (1996) Benchmarking, best practice and all that. *Third International Conference on Lean Construction*. University of New Mexico Albuquerque, 18–20 October.
25. McCabe, S. (2001) *Benchmarking in Construction*. Blackwell Science, Oxford.
26. Latham, M. (1994) Constructing the team: HMSO, London.
27. Egan, J. (1998) Rethinking construction, DETR, London.
28. Construction Best Practice Programme (CBPP) web site (2002) http://www.cbpp.org.uk/cbpp/index.jsp
29. Yoshimori, M. (1995) Whose company is it? The concept of the corporation in Japan and the West. *Long Range Planning*, **28** (4), 33–4.

Chapter 5
Reengineering

Introduction

Despite being a relatively new discipline there is already some ambiguity with respect to what is meant by 'reengineering'. For a start there are variations in the spelling. Some commentators favour the hyphenated version 're-engineering', others use the non-hyphenated version 'reengineering'. Hammer, who is usually credited with the introduction of the concept, favours the non-hyphenated version and we have opted for this spelling in this text. Most general use dictionaries, such as Webster's, do not recognise the term in either form.

Confusion can also arise on whether to use 'reengineering' as the generic term which applies to all reengineering activities or to use the more specific term 'business process reengineering' (BPR). Because most references on the subject are from a business background, the term BPR is frequently used as the generic. We have however opted to use reengineering as the generic term (as does Hammer) because the term construction process reengineering (CPR) has recently been coined[1] and, for the purposes of this text, we need to be able to distinguish between BPR and CPR.

Much is currently being claimed about the power of reengineering. The flyleaf of Hammer and Champy's text *Reengineering the Corporation*[2], makes the claim that

> 'In *Reengineering the Corporation*, Michael Hammer and James Champy do for modern business what Adam Smith did for the

industrial revolution in the *Wealth of Nations* two centuries ago – they reinvent the nature of work to create the single best hope for the competitive turnaround of business.'

As we have stated in the preface, the purpose of this book is not to champion any particular management concept in preference to another. We have throughout the text attempted to be as detached and objective as possible. However, of all of the concepts in this book, reengineering has been the most difficult for us to maintain our level of detachment. This is, in the main, because of the fact that reengineering is seen by many of its proponents as a revolutionary rather than an evolutionary concept, so much of the language used to describe reengineering has an emotive flavour. For example, Hammer's often-quoted saying 'don't automate, obliterate', typifies the revolutionary hype of reengineering parlance. Many texts on the subject make claims such as 'radical change', 'dramatic results'. Hammer and Champy use both these expressions in defining reengineering as 'a radical redesign of business processes in order to achieve dramatic improvements in their performance'. Morris and Brandon[3] take the extreme view that 'You can choose to reengineer or you can choose to go out of business.' It has been difficult for us to present a cool clinical exposition of the topic in the light of comments such as those by Morris and Brandon. There is however a counterpoint to this aggressive promotion of reengineering in the work of COBRA[4] (Constraints and Opportunities in Business Restructuring - an Analysis) which is an initiative of the European Commission. Professor Coulson-Thomas, leader and co-ordinator of the COBRA project, describes the backdrop to the project as follows:

> 'Creativity and imagination are at a premium. According to its theory and rhetoric Business Process Re-engineering (BPR) is concerned with step change rather than incremental improvement; revolution not evolution. However, while BPR is being used to "improve" existing situations, i.e. cut costs, reduce throughput times and squeeze more out of people, what does it contribute to radical change and innovation?'

What we have tried to do for the reader is to take the middle ground, as far as this is possible, in presenting an overview of reengineering and exploring the implications of reengineering for the construction industry.

We should say at the outset that we are more inclined to Coulson-Thomas's view that reengineering is not the only way to achieve radical and fundamental change and that creative thinking, benchmarking, cultural change and innovation can be undertaken independently of reengineering. Reengineering is however an important management concept which has achieved quite dramatic and tangible improvements for large business corporations and, as such, merits serious consideration.

Origins of reengineering

Although the emergence of reengineering is usually attributed to the publication in 1990 of Hammer's *Harvard Business Review* article[5] 'Reengineering work: Don't automate, obliterate', the origins of the reengineering concept can be traced back to the 1940s to the work of the Tavistock Institute in the UK where the 'social technical systems approach' was applied to the British coal industry[6]. The essence of the social technical systems approach (STS) is that technical and social systems should harmonise to achieve an optimal overall system. The STS approach was translated in the 1980s[7] into Overhead Value Analysis (OVA) which emphasises that attention must be centred on the work being done rather than on the people doing the work. According to Rigby[6], OVA was in fact the precursor to reengineering.

The underlying theme of the STS, the OVA and the reengineering approach is that work processes need to be redesigned. The implications of reengineering, according to Hammer, are not that we design inefficient processes to begin with but that these processes have been overtaken by time and have not evolved or kept pace with advances in technology.

> 'Over time, corporations have developed elaborate ways to process work. Nobody has ever stepped back and taken a look at the entire system. Today, if most companies were starting from scratch, they would invent themselves in totally different ways.'

The added dimension which reengineering gives to STS and OVA is to the concept of 'discontinuous improvement'. The imperative for reengineering is to achieve a quantum leap forward rather than small continuous gains[8].

Reengineering in a construction industry context

In considering reengineering in a construction industry context, the first thing which needs to be determined is whether or not the construction industry is a special case, i.e. should it be treated differently from all business processes? There are arguments for and against treating construction organisations as somehow different from other business organisations. Certainly the construction industry has a set of characteristics and a product which is uniquely its own. However delivering a product with a unique set of characteristics is not a convincing argument for special status. The automotive industry and the white goods industries produce quite different products, one producing cars which must be able to perform in the open air and the other domestic appliances which operate in an internal environment. However no one would suggest they are not both part of the business process.

The argument is often made that the construction industry is highly compartmentalised, highly fragmented, under-capitalised, and operates on a single project-by-project basis, and so on. The more that we examine the special pleadings that the construction industry is different from all other business processes, the more convincing becomes the argument that the industry is a prime candidate to be reengineered. To advance this argument further, the starting-off point for the reengineering process is to take a new or green-field approach, a feeling of wiping the slate clean. There is no better way for the construction industry to do this than to regard itself as a member of the international business community, on a similar footing to the McDonald's, Xerox's and Toyota's of the world. Although, in general, the scale of the construction organisations is smaller than that of other multi-national business organisations, the output of the industry is usually in the order of 8 to 10% of the Gross Domestic Product (GDP) of a Westernised economy, with a 10%

improvement in construction performance representing a 2.5% increase in GDP.

The view which we would like readers to consider, in the context of reengineering, is that the construction industry should be seen as *part of*, not *apart from* the general business community. This is not meant to infer that the construction industry has everything to learn and nothing to contribute to the business community at large.

The current trend in business process reengineering applications is for these to be internally focused within the organisation:

> 'the BPR application becomes exclusively an "in-house" operation, in which, decisions pertaining to meeting customer requirements, investing in technology and organising resources are steered and, if necessary, manipulated by the organisation, with minimum dependence upon and interference with external factors[9].'

Whether or not reengineering can be applied to an industry sector as a whole rather than to individual organisations is a challenge which has still to be addressed.

> 'An industry, such as the construction industry, represents a major challenge to BPR. This is mainly due to the complex business relationships which dominate the industry plus the key role external factors play in how construction industry organisations conduct their business[9].'

Our point is that although the construction industry does have peculiar characteristics, it is still part of the business community. If the construction industry can apply reengineering not only at the level of the firm but also at the level of a complete industry sector, this will be a major contribution to the advancement of the reengineering cause and would be a considerable achievement for the construction industry. Whether or not this can be achieved is another matter, which will be considered later.

The goals of reengineering

For many years the construction industry has focused on delivering buildings on time, within budget and to a specified quality. These goals are quite compatible with the goals of reengineering, which are 'to achieve dramatic improvements in critical contemporary measures of performance, such as cost, quality, service and speed'[2], an area that the construction industry is only recently coming to grips with. By and large the goals of reengineering are no novelty to the construction industry. What is novel however is the green-field methodology which is the essence of the reengineering concept. Hammer and Champy[2] distil reengineering into four key words or characteristics: fundamental; radical; dramatic and process. The implications of these key words pose some interesting questions for the construction industry.

The first characteristic of the reengineering approach is the requirement to take a fundamental look at what is the core business of a company. In essence the questions being asked are 'why are we here?', 'why do we operate the way that we do?' The expected response from a business organisation to the question 'why are we here?' would be 'to service our customers' or 'to meet our customers' expectations'. Regrettably the construction industry does not have a good track record with respect to identifying the customer as the *raison d'être* for the organisation's existence. A recent survey of construction companies in Australia demonstrated that although many companies profess an interest in obtaining feedback on customer involvement, only 5% of those surveyed conducted a formal survey to gauge satisfaction levels[10]. Reengineering demands that nothing be taken for granted, particularly the client. The status quo is not sacrosanct. The primary reengineering objective, when applied to the construction industry as a business process, is to exert pressure on the players in the industry to truly understand the nature of their business.

The second characteristic of reengineering is its emphasis on taking a radical approach. The word 'radical' is often taken to mean 'new' or 'novel'. It is worth remembering that the derivation of radical is from the Latin word *radix* meaning root. *Webster's Dictionary* defines radical as ' relating to the origin', 'marked by a considerable departure from the usual or traditional'. (Hammer[2] makes particular emphasis of this.) A radical approach in the construction industry would be to ask the question 'how can we redesign the process so that waste is eliminated?' rather than 'how can we minimise materials wastage on site?' Or at a macro level, consideration could be given to using the principles and tools of 'lean construction' which encompass the whole gamut of decreasing waste in both the design and construction process.

It is worth stressing that although reengineering demands a radical approach to problem definition and problem solving, this does not mean that existing management tools should not be deployed. Although the essence of reengineering is the novelty of its approach, the use of management tools such as total quality management, benchmarking, concurrent construction and lean construction are essential components of the reengineering approach.

There seems to be general agreement that Hammer and Champy's third characteristic, 'dramatic', typifies the successful application of reengineering. The effects of a successful reengineering strategy should be discontinuous improvement, i.e. a quantum leap forward. The notion of a quantum leap may seem to infer that only companies who are performing badly will reap the full benefits of reengineering. Hammer and Champy dispute this and identify three kinds of companies that can benefit from reengineering. These are companies who are in deep trouble and have no choice but to undertake reengineering in order to survive. The second company type is one which is not yet in trouble but whose management has the foresight to see that trouble is brewing ahead. The third company type is one which is a leader in its field which is not under threat either now or in the foreseeable future but is keen and aggressive. Thus no sector of the business community is absolved from involvement in the dramatic impact of reengineering.

One question which does however spring to mind in terms of the dramatics of reengineering is 'how can a company exist in a state of perpetual discontinuous improvement?' In other words once having achieved a quantum leap in business performance through reengineering, will the law of diminishing returns set in so that additional applications of reengineering result in ever-decreasing gains? This conundrum has been addressed by Morris and Brandon[3] in what they describe as the 'change paradigm'.

> 'The change paradigm is a conceptual environment. Once a company begins operating within this paradigm, the process of reengineering never ceases. It becomes constant but incremental, as the company evolves toward better quality and efficiency. This represents a new business operation cycle.'

Thus the change paradigm addresses the conundrum of the potentially self-defeating nature of reengineering by recognising that in the future the construction industry will compete in an environment which is in a dynamic state of change where the only constant is change itself.

The fourth and final key word in Hammer and Champy's definition of reengineering is 'process'. The underlying contention is that most businesses are not process-oriented. The primary concern of reengineering is in redesigning a process, not in redesigning the organisational structure of departments or units. The distinguishing characteristic which reengineering brings to this redesign is that it should be fundamental, radical and dramatic.

> 'The majority of the organisations who use the term business process reengineering are in reality engaged in process redesign. For most companies this approach represents radical change. However, it is not what Michael Hammer and James Champy meant by the word "radical"[11].'

It is important to differentiate between process improvement, process redesign and process reengineering (Figure 5.1)[11].

Process improvement involves a minor degree of change with a

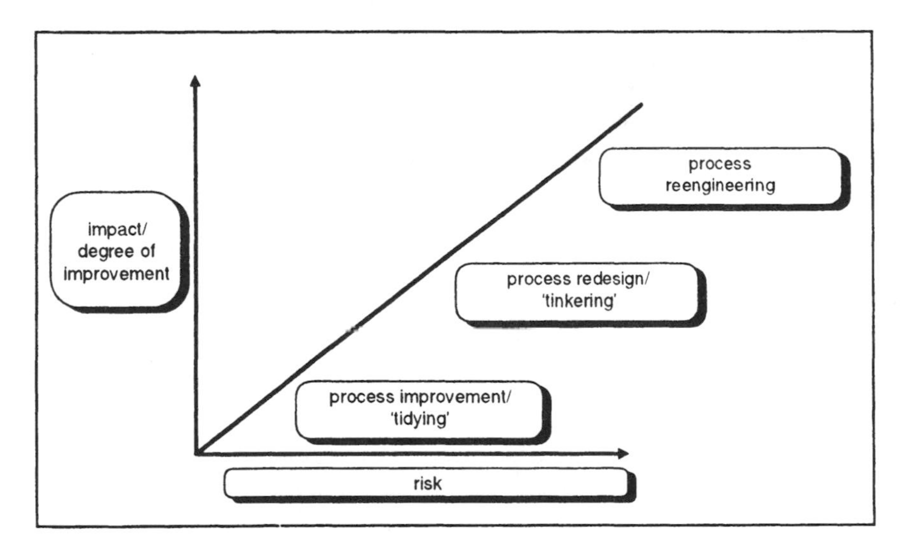

Fig. 5.1 Differentiating between process improvement, process redesign and process reengineering.

corresponding low degree of risk together with a low expectation of improved results. Process redesign is the middle ground with a moderate degree of risk and consequently higher expectations of improvement. Process reengineering is at the end of the spectrum with a high risk, high gain set of expectations.

These three levels have alternatively been described as process tidying, process tinkering and process reengineering[12]. Process tidying is described as a method by which existing flows of people, information and materials are mapped and streamlined by identifying opportunities for eliminating dead-ends, ad hoc activities and duplication. Process tinkering is described as a method by which organisations find short-cuts in their processes or identify more user-friendly ways of doing work. In general, process tinkering does not seek to change the overall process nor does it seek to move constraints, whereas process reengineering is the method by which physical or mental recipe-induced constraints are eliminated from the organisation and re-established in a way which meets the goals of the organisation.

Reengineering methodology

The initiation stage

Reengineering is about cultural change. It is concerned with decisions of a strategic and often political nature. It is about changing attitudes and behaviour and points of reference. It follows therefore that tactically, reengineering must be initiated at the highest level within a company. It is a top-down approach. Hammer and Champy[2] observe that the following roles tend to emerge in companies which have implemented reengineering:

The leader
In a construction company this would typically be the managing director or someone with the equivalent authority who would initiate the reengineering concept.

The process owner
Reengineering will be initiated at head-office level, not at site or

project level. The process owner is therefore likely to be a director or very senior manager based at head office.

The reengineering team

The team would consist of a multi-disciplinary group who are committed to the concept of reengineering with skills and competencies ranging from construction management to IT.

Steering committee

This would be a small group of senior managers who are highly regarded within the construction company.

Reengineering 'czar'

In Hammer and Champy's words 'an individual responsible for developing reengineering techniques and tools within the company and for achieving synergy across the company's separate reengineering projects'.

DuBrin[13] places particular emphasis on the importance of the process-owner or case-manager as a person and also the importance of this person having the authority to leap over departmental walls.

The examples of reengineering very largely emanate from a factory floor production environment, which has organisational groupings which are not necessarily akin to the construction industry. It is worth bearing this in mind and, while there is nothing in Hammer and Champy's definition of reengineering roles which would be an anathema to the construction industry, we must be aware that in the construction industry we are dealing with a bespoke product derived from an organisational structure which usually has a complete separation of the client, designers, constructors and users. Thus the process-owner in the construction industry reengineering context must, if we use DuBrin's analogy, have the ability to leap over a large number of extremely high walls.

The planning stage

There seems to be a general agreement among commentators and practitioners of reengineering that the benefits of reengineering come from the holistic nature of the concept. In reengineering the emphasis is on horizontal rather than vertical integration, with the breaking down of barriers, the removal of specialist functions and basically a redrawing of existing boundaries. The overbridging strategic methodology recommended by most practitioners has its origins in the systems approach.

The conventional systems approach to planning a project is first to identify the system in general terms by establishing the system boundaries and then, and only then, proceed to the specifics. This approach is illustrated in Table 5.1 after Armstrong[14]:

Table 5.1 The conventional systems approach.

Conceptual	Operational
(Start here) Ultimate objectives	Indicators of success
Alternative strategies	Operational programme *(Finish here)*

In a classical systems approach each step should have a separate time period with time for reflection between each step. Each step should be considered without reference to the steps which come later. The sequence of steps is then applied iteratively.

The application of systems thinking can be seen in Table 5.2 which is loosely based on the Texas Instruments diagram of the components of business process reengineering[11].

Step one 'identify objectives' means starting at the highest conceptual level and determining 'what the business needs to be' by resolving the stakeholder interests of 'what the customer wants' and 'what are the business objectives?'. This first step should be com-

Table 5.2 The application of systems thinking.

Steps	Procedures
Identify objectives	Resolve customer needs and business objectives
Develop indicators of success	Indicators of success are developed through an understanding of the business process
Consider alternative strategies	Select either: business process reengineering or continuous improvement
Develop and select programmes	If BPR selected then embark on a programme of fundamental change. If continuous improvement selected then embark on a programme of fine tuning

pletely resolved before proceeding to the next step, 'develop indicators of success'. Developing indicators of success is an important element of the systems approach; this can only be done through a thorough understanding of what the 'business is'. Once the first two steps have been completed, decisions can then be made in terms of selecting the alternative strategies of either business process reengineering or continuous improvement, thus leading to step four, the development of programmes. The whole process is dynamic with the steps being repeated, in sequence, as required.

The COBRA project[15] developed a six-stage business process reengineering methodology which has similar characteristics to the Texas Instruments model. It also includes the iterative loop of the systems approach.

The implementation stage

As we will discuss later in this chapter, there are pitfalls associated with the implementation of reengineering. However, for the moment we will concentrate on the factors which are likely to contribute to successful implementation. It is worth however bearing in mind the contention 'that radical change is accompanied by substantial risk'[11]. In reengineering the risks are heightened by the length of time needed to complete the implementation stage and to get results. The need to 'get on with it' must always to be tempered by the need to prove, on a step-by-step basis, that the implementation process is working.

Coulson-Thomas[16] identifies 15 success factors in implementing reengineering based on his own experience and observations. Many of these factors relate to the 'people' component of reengineering. In his view reengineering is, in essence, the effective management of fundamental change. He cites success factors such as mutual trust between senior managers and change teams as being critically important: in other words, reengineering is largely concerned with 'feelings, attitudes, values, behaviour, commitment and personal qualities such as being open minded'.

The importance of the human/communication factor is also reflected in the following key success factors identified by Mohamed and Yates[17] as prerequisites in implementing reengineering in a construction industry context. (Readers with a particular interest in how reengineering might be applied to a specific construction work package are recommended to read this paper in full.)

- ❑ Strong commitment by designers, consultants, and contractors to make a major shift in the way the existing workflow structure of design and construction works.
- ❑ An effective communication cycle must be established and maintained between major project participants as information exchange helps eliminate rework and consequently reduces time.
- ❑ Positive involvement of external as well as internal customers must be sought at the early project stages so that their requirements can be captured and implemented as input in the planning stage.
- ❑ Quality assurance techniques must be developed and implemented across the various work elements of the construction process.
- ❑ Innovations should be encouraged in areas of planning, contracting, design and construction.
- ❑ New approaches should be investigated to improve construction output.

Whereas all commentators are in agreement that the success or failure of reengineering hinges on the achievement of a cultural shift, there is less agreement on the detailed mechanism for bringing this about. Petrozzo and Stepper[18], for example, disagree with Hammer and Champy 'that the leader of the reengineering project does not have to be involved in a near full-time capacity'. There are also differing views on the appropriate management style to be adopted. Petrozzo and Stepper[18] argue that

> 'the reengineering leader must *aggressively* communicate the importance of the reengineering effort, what it is going to do, and when it will be implemented' (our italics).

To some, aggressive communication would seem to be a contradiction in terms and somewhat at variance with Coulson-Thomas's more spiritual view that

> 'organisations are living communities of people. They are sensitive organisms, reflecting our dreams and fears. Knowledge workers are increasingly attracted to those networks whose values they share'[4].

Because of the philosophical and cultural differences which exist between, say, European and North American corporations, it is not possible for us to give an explicit method of reengineering which will hold good for all situations. If we take the case in point of Petrozzo and Stepper versus Coulson-Thomas, then some employees may respond positively to a hard-sell dictatorial approach which in other organisations would be a recipe for disaster. It is perhaps easier to say what the results of the reengineering process look like rather than to attempt to describe a step-by-step progression through the implementation process. Even here however, Hammer and Champy[2] state that is not possible to give a single answer to 'what does a reengineering business process look like?' They do agree however that, not withstanding this caveat, there are a number of points of agreement which, by and large, typify the results of implementing reengineering. These are:

- ❑ several jobs are combined into one
- ❑ workers make decisions
- ❑ the steps in the process are performed in a natural order
- ❑ processes have multiple versions
- ❑ work is performed where it makes the most sense
- ❑ checks and controls are reduced
- ❑ reconciliation is minimised
- ❑ a case manager provides a single point of contact
- ❑ hybrid centralised/decentralised operations are prevalent.

The explanation of the above set of characteristics is given in the context of North American business corporations and production processes which are taken from the factory floor.

In the construction-industry context the following potential time and cost savings which would result from the application of reengineering have been identified by Mohamed and Tucker[9]. (This paper is of particular interest because it considers BPR at an industry level (in this case the construction industry) as opposed to the more common focus of BPR at the level of the firm. Readers are recommended to read this paper in full.)

Potential time saving

- Reducing the extent of variations and design modifications as a result of the systematic consideration of customer requirements in a clearly and accurately developed design brief.
- Producing less timely and more effective solutions while meeting client requirements by involving other engineering disciplines in the design phase.
- Achieving an agreed competitive overall construction time by negotiating better ways of construction with a contractor, who is selected on the basis of recognised past performance and financial stability.
- Improving the quality of design by incorporating the contractor's input in the design phase. This is conducive to smoother site operations and fewer construction delays.
- Compressing overall construction time by considering constraints imposed by downstream operations such as building approvals, material availability and site conditions during the design phase.
- Increasing the productivity of subcontractors by reorganising small work packages into larger ones, thus minimising delays caused by poor co-ordination and interference.
- Improving project performance by enhancing the working relationship between project participants through the adoption of team-building, partnering and strategic alliance concepts. These concepts minimise the possibility of construction delays caused by conflict of interests.
- Increasing the efficiency of construction performance by reducing or eliminating inherent time waste in project material and information flows.

Potential cost savings

The majority of the above listed areas of potential time savings can lead, either directly or indirectly, to cost savings. In addition to the cost benefits attained because of the reduction of the overall construction time, other main potential areas of cost savings can be summarised as follows:

- Developing a design brief that accurately represents client requirements safeguards against additional costs caused by design modifications, omissions and associated delays.
- Selecting a design option through the application of the value management concept implies avoiding more costly options for the same functional needs (clients get best value for investment).
- Applying the concept of concurrent engineering and considering the downstream phases of the constructed facility helps the client select a design option with lower operating, maintenance and replacement costs (life cycle costs).
- Appointing a contractor based upon both past performance and financial stability reinforces to a large degree the process of controlling financial and operational risks.
- Implementing the concept of team-building between key project participants and partnering between the contractor and subcontractors enhances working relationships and reduces the number of costly conflicts and claims.
- Adopting proper quality measures into the design and construction processes ensures minimising the amount of rework for which the client ultimately pays.
- Employing an efficient material management system, such as the (Just-In-Time) approach, saves operating costs associated with material handling, storage, theft, damage, etc.
- Having project data communicated in a more timely and accurate manner reduces the possibility of extra costs resulting from decisions based upon poor or outdated information.

Pitfalls of reengineering

At the heart of reengineering lies the systems approach, with its holistic perspective. One cornerstone of systems theory is that if one attempts to optimise the components of a system in isolation from one another, the inevitable outcome will be a suboptimal solution for the system as a whole.

> 'Undertaking self-contained BPR exercises at the level of the individual process can actually reduce the prospects of wider transformation. As a consequence of making an existing form of organisation more effective, the impetus and desire for an overall transformation may be reduced[4].'

Because reengineering is very largely concerned with the breaking down of conventional subdivisions, for example, the reengineering case manager who has the power to leap departmental walls, then it is inevitable that the relationship of subsystems to the system as a whole will be a continuing preoccupation of reengineering practitioners. An instance of the suboptimal tendencies of BPR is where a BPR empowerment drive which encourages and trusts people to be flexible and catholic in how they work can be in conflict with a parallel BPR initiative seeking to define particular ways of approaching certain tasks. Or take a further instance where market testing can result in the carving up of an organisation into a collection of contractual agreements of varying timescales with the result that, in effect, there is no longer an organisational whole to reengineer or transform[16]. The natural reaction in the systems approach to preventing suboptimal solutions is simply to keep widening the system boundaries in order to be all-encompassing. Increasing the systems boundaries is not however always a practical, or even a conceptually desirable condition.

This is not to infer that there is an inherent conceptual flaw in the reengineering philosophy, but simply to point out that the self-induced challenges of reengineering are substantial, not to say daunting. Most commentators would agree that it is a high risk and high gain occupation[11].

At an operational level Hammer and Champy list the following as mistakes to be avoided in the application of reengineering.

- Trying to fix a process instead of changing it.
- Not focusing on business processes.
- Ignoring everything except process redesign.
- Neglecting peoples' values and beliefs.
- Being willing to settle for minor results.
- Quitting too early.
- Allowing existing corporate cultures and management attitudes to prevent reengineering from getting started.
- Trying to make reengineering happen from the bottom up.
- Assigning someone who doesn't understand reengineering to lead the effort.
- Skimping on resources devoted to reengineering.
- Burying reengineering in the middle of the corporate agenda.
- Dissipating energy across too many reengineering projects.
- Attempting to reengineer when the CEO is two years away from retirement.
- Failing to distinguish reengineering from other business improvement programmes.
- Concentrating exclusively on design.
- Trying to make reengineering happen without making anybody unhappy.
- Pulling back when people resist making reengineering changes.
- Dragging the effort out.

While the list is an informative set of instructions in its own right, it is also a useful insight into the pitfalls which can occur in trying to introduce reengineering into an organisation and also gives a very clear view of the traps that exist for the unwary.

A case study on the introduction of reengineering into a multinational electronic component company in a project named 'Smart Moves' gives the following lessons learned[19]:

> 'Firstly reengineering is not an easy or automatic activity. If it is to take place it cannot be done half-heartedly or in half measures. Although the methodology of selecting the process to be reengineered is not difficult to put in place, the execution of this is not an easy matter. Firstly, an appropriate process must be selected, capable of adding values and which can be clearly and concisely defined. The team must be empowered and an executive cham-

> pion found. Additionally, getting a thorough understanding of the current process and representing this clearly and unambiguously can be tedious. Moreover developing a vision of the reengineered process through to the negotiating and executing the plan can be fraught with difficulty. A radical change may be identified by the team but this may be very difficult for management to commit to a high risk project, especially one which alters the status quo.'

As a counterpoint to these perceived difficulties the authors note that the reengineering process has serendipitous effects which are difficult to define but none the less exist. They conclude by remarking that 'Even if the radical changes required by BPR are not feasible, business process enrichment may enable existing processes to add value whenever possible.' A concluding remark which seems to be somewhat rueful.

Information technology and reengineering

Just as reengineering is seen as a means of empowering modern business to effect radical change, so information technology (IT) is seen as the enabling mechanism to allow radical change to be effected. Many believe that reengineering is intrinsically linked to the application of IT[20], although

> 'BPR does not absolutely require the use of IT at all. However, one of the distinctive features of BPR is the way that IT is almost invariably used to "informate" in Zuboff's terminology, to reconstruct the nature of work. It is a key feature of implementing BPR[21].'

There is a 'push-pull' synergistic relationship between reengineering and IT, with the power of IT extending the horizons of reengineering and the challenges of reengineering acting as a catalyst for new IT development.

IT can be extended to ICT (information and communication technologies)[22] and this is a useful broadening of the scope of IT, particularly when considering its relationship to reengineering.

Several commentators[21,22] refer to the Venkatraman model[23] with respect to the relationship between IT/ICT developments and the business process.

The five stages of the Venkatraman model (figure 5.2) have been described as follows[22]:

The first level of the model is termed 'localisation exploitation', and involves the employment of standard IT applications with minimal changes to the organisational structure of the firm. This level represents discrete 'Islands of Automation' within the business process.

The second level, 'internal integration' can be thought of as first building the internal electronic infrastructure which allows the integration of various tasks, processes and functions within an organisation and then uses this platform to integrate these intra-organisational processes. Venkatraman argues that this is an evolutionary step from 'localised exploitation'.

The third, and first of the revolutionary levels, business process redesign, uses ICT as a lever for designing new business procedures rather than simply overlaying the technology on the existing organisational framework.

The fourth level, business network redesign, involves the use of ICT to step beyond traditional intra-organisational boundaries to include clients, suppliers, and changes of the competitive environ-

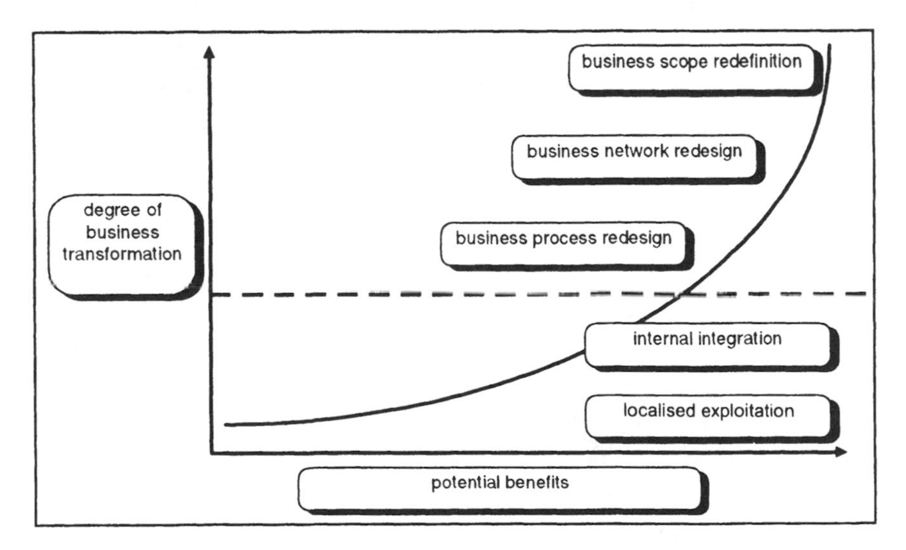

Fig. 5.2 Five levels of itemised business.

ment by introducing a need for greater co-operation between trading partners.

The final level, business scope redefinition, is the end point to the model in which Hazlehurst[22] suggests the question 'What role, if any, does ICT have to play in influencing business scope and the logic of business relationships within the extended business network?'

The Venkatraman model is useful in categorising levels of IT-enabled business particularly with respect to the threshold between evolutionary and revolutionary levels. Holtham[21], rightly, stresses the need to distinguish between the term 'process' as applied to business process and the term 'process' as it is applied in the field of computer science and software engineering. He makes the point that some IT advocates have used level four of the Venkatraman model as a justification for merely upgrading or replacing existing computer systems, which, of course, is at variance with the thrust of the Venkatraman model and with the philosophy of reengineering.

There are very few examples of the implementation of reengineering in the construction industry, but this is not to say however that there are no examples to hand of IT or ICT-enabling technology which could be deployed in reengineering activities. Although the advancement in computer technology and the widespread use of computers in the industry have opened a new frontier in construction applications, applications are still far from being satisfactory because of historical, financial and operational reasons. Notwithstanding this caveat, Mohamed and Tucker[9] have identified the following IT technologies as enabling reengineering in the construction industry:

- CAD conferencing which allows project participants, e.g. the client design team and contractors, working from different geographical locations and different hardware and CAD systems, to work collaboratively. Future developments in this area will link design offices and overseas construction sites.[9,24]
- The use of virtual reality for planning and construction activities.
- Relational data base shells for use in a range of building procurement activities.
- Knowledge-based systems for use in building codes linked to standard exchange for product data (STEP).
- Computer-based simulation of workflow leading to the application of automated and semi-automated robots.

- Automated cost engineering using modular, computerised knowledge bases accessed though a graphical user interface.
- The development of organisational decision-support systems using a graphical interface.

The above examples of leading edge IT applications are focused, in the main, on the planning and development stage. The following examples relate to the simulation of construction site operations[25]:

- statistical process simulation of construction operations
- the use of graphical simulation to create a virtual site environment
- the use of simulation in the application of 'lean site' operations encouraging the use of resource minimisation techniques such as just-in-time.

Care has to be taken that because of the nature of the technology, IT has a seductiveness which may in fact detract from, rather than enable, the reengineering effort[26]. While many companies have used IT in the furtherance of successful reengineering,

> 'Equally, there are a number of examples of organisations that used little or no new investment in IT to achieve their reengineering goals. In such cases the aim is to:
>
> - Take IT off the critical path.
> - Make use of existing systems as far as possible.
> - Use flexible front ends to provide the appearance of seamless common user access to a range of underlying systems.
> - Only redevelop once the process has been redesigned and the new IT requirements, if any, are clear to all involved.'

This seems sensible advice in preventing the misuse or wrongful application of IT in reengineering. One of the classical problems in computer applications, which existed long before the introduction of the reengineering concept, is whether to use the computer to automate the existing manual system or whether to take the opportunity presented by a new computer technology to rethink the existing process. The tendency has been first to automate the existing procedure, then at a later stage to recognise that an opportunity has been lost to really harness the power of the computer, and then to backtrack to radically rethink the process. The golden rule as far as IT is concerned is only redevelop once the process has been redesigned, and the new IT requirements, if any, are clear to all involved. IT has a powerful role to play as an enabler, but not an instigator, of reengineering.

Reengineering from a European perspective

Reengineering has much to do with cultural values. It is, by its very nature, an emotive concept. It has been our intention to try to steer a middle course in presenting the current reengineering movement without letting our own personal views intrude. We have included reengineering in this book because we believe that it is an important modern management movement and one worthy of serious consideration by the construction industry. However, it would be remiss of us if we were not to include some of the quite severe reservations which are held by some senior managers in Europe in terms of the impact and potential of reengineering.

The head of Siemens, Heirich von Pierer, writing on Business Process Reengineering, has been quoted[21] to the effect that 'I don't feel completely comfortable with the radical thesis of Mr Hammer. Our employees are not neutrons, but people. That's why dialogue is important.' Holtham[21] goes on to express the view that BPR needs to be rooted in distinctive European managerial features, in terms of the

> 'acceptance of the humanistic and holistic stream in European thought, in contrast to the more mechanistic and fragmented US approach with the promotion of the concept of collaboration, between levels in the organisation, across organisations, between supplier and customer, and also across national boundaries. There would appear to be little variance between Holtham's vision of BPR in Europe and Hammer's description of the application of BPR in the States. Perhaps it is more a disagreement on the method of delivery than the message being delivered.'

As Holtham concludes, 'The core elements of BPR have value

beyond the evangelical North American approach, and BPR has value for Europe, if it is set in a European context.'

A case study of a process reengineering study in the Australian construction industry

There is no single answer to 'what does a reengineering business process look like?' Given that the very nature of the approach is dependent on creative thinking and the breaking down of existing structures and barriers, it is not possible to have a specific procedure or checklisting approach on how to implement reengineering. The following case study however does capture many of the features of reengineering through the initiation, planning and implementation stages.

(The text of this case study has been extracted by the authors from the final T40 report. We are indebted to the T40 contributing organisations for giving their permission to allow us to reproduce this material. For a more detailed discussion of the T40 project readers are referred to Ireland[27].)

Background

The T40 research project[27] was a process reengineering study in the Australian construction industry whose objective was the reduction of construction process time by 40%.

The participating organisations are shown in Table 5.3. This grouping represents three major contractors, two major services suppliers, two key consultants and CSIRO (the Australian national research and development organisation). The study was conducted in 1993/94 and reported in May 1994.

Table 5.3 Participating organisations.

Fletcher Construction Aus.	CSIRO	BHP Steel
CSR	A.W. Edwards	Stuart Bros.
James Hardie Industries	Otis	Smith Jesses Payne Hunt
Taylor Thomson Whiting	Sly & Weigall	Motorola (USA) facilitators

Research method

The T40 team was led by Professor Vernon Ireland of Fletcher Construction Australia, with each of the contributing companies having responsibility for an aspect of the process. Funding of over Aus. $300,000 was provided with the largest portion ($96,000) coming from the Australian Building Research Grants Committee.

The T40 group engaged Motorola (USA) to facilitate the analysis of the construction process as a process reengineering exercise, drawing on Motorola's experience in process reengineering their own manufacturing and project-management operations. Motorola were also able to provide a non-construction industry perspective, which had no allegiance to the conventions of the construction industry.

The T40 study had three focal weeks of workshops which explored:

- ❑ Flow charting the 'As-is' process
- ❑ Redesigning the 'Should-be' process
- ❑ Developing aspects of the solution.

All members of the team contributed to the final document. The document reflects the experience of team members, who represent most of the participants in the design and construction process. The intention was to produce a redesigned process and a series of new practices rather than simply complete present practices faster. Some of the proposals can be implemented with little disruption to current practices while some will require extensive discussion with other key parties in the industry before implementation.

Themes for the T40 project

The research findings included a number of key themes:

- ❑ Organisation of the T40 solutions team into a small group of up to nine organisations, rather than a large group of general contractor plus specialists (50–100), with each T40 team member addressing the clients' needs directly.

- Reorganisation of work packages to eliminate multi-visits to the same construction location by the same specialist.
- Maintenance of single point accountability for the client.
- Financial incentives and penalties for the whole group of nine (including architect and structural engineer) to focus action.
- Business practices based on trust and fair dealing, thus eliminating the checkers, who are checking on other checkers on other organisations.
- Getting it right first time with the elimination of rework.
- Elimination of traditional tendering, thus eliminating the time and cost of tendering as well as allowing the solutions team to directly answer the customer's needs.
- The solutions group sharing resources rather than duplicating functions (e.g. planning, supervision, employment of licensed trades and administration).
- Teaming between management and the workforce.
- Partnering with local government for approvals.

The T40 team recognised that changing the attitudes of key participants to achieve the innovations will not be easy. Many people in the industry will say that the goals are unachievable. However the team believes that such changes are necessary for the development of the Australian industry and it is feasible to achieve the goals over a period of time. The achievement of such goals will put Australia in a key position to take a strategic role in assisting the large industries in Asia.

Proposed T40 process

Agreed common goal between customer and delivery team

The T40 proposal is that all members of the solutions team will directly address the customer's needs, rather than have them filtered through the architect and filtered again through the general contractor. The whole team will directly focus on providing a solution which adds value to the customer's business.

Both the customer and the solutions team members need to state what they want from the process and the limitations to satisfying the

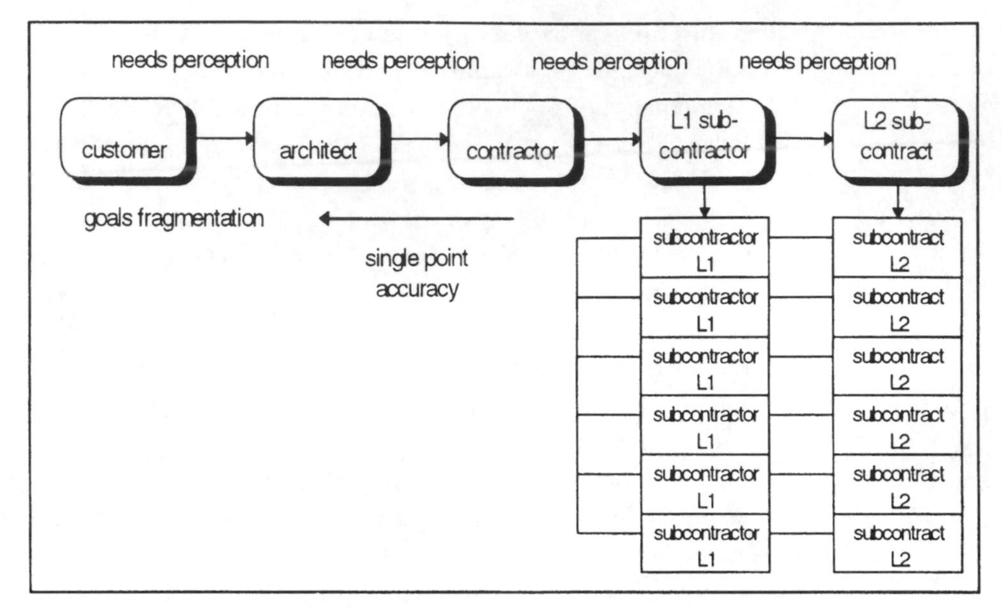

Fig. 5.3 Traditional process hierarchy.

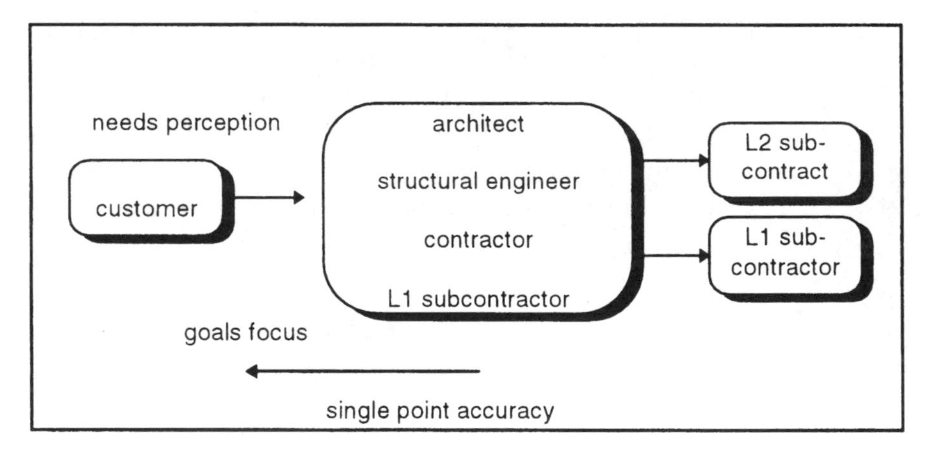

Fig. 5.4 T40 solutions.

goals of the other. To achieve such an approach the following changes to the status quo are needed:

- New methods of dependence and trust must be evolved, which will be enhanced by hunting for projects as a team together with financial incentives to support team play (shared success bonuses and penalties).

- New work patterns to reflect the dependence must be developed, for example the acceptance of a reallocation of traditional work packaging.
- Commitment must be obtained from the client and participant team at the highest corporate level.

Simplified process

The proposed T40 project delivery system is a one-stage design and construction process with the whole solutions team involved from the point of determining the customer's needs. The key issue is that the customer is satisfied that value for money is being achieved and the project is delivered in the agreed reduced time.

It is a fundamental belief of the T40 research team that a construction team of 50–100 separate contractors is not as efficient in the long term as a group of eight to ten key players. Project managers with spans of control of 50–100 subcontractors and suppliers can only fight fires and have limited scope for innovations.

The T40 proposal is that, in a typical high-rise office development, one contractor would be responsible for each of the following:

- Site establishment }
- Excavation } lead
- Structure } contractor
- Enclosure }
- All services except lifts (design and construction)
- Lifts (design and construction)
- Landlord tenant fitout
- Concrete supply
- Steel supply

The architect and structural designer are also full members of the team, with ownership of goals and taking an appropriate share of risk and reward.

During the fitout phase there would be two services associates and two fitout associates. While this arrangement does introduce a three-layer hierarchy for the fitout and services contracts, the traditional reasons for there being no scope for control and performance incentives for these specialists can be overcome by separation in time and location.

It is essential that there be a clear definition of work packages between associates to eliminate interference and allow risks and rewards to be adequately apportioned. Innovation that will occur within the package subcontracts will include:

- Development of innovative physical systems.
- Better integration of activities between the third layer of subcontractors.
- Reengineering the subprocesses to eliminate co-ordination problems.
- Analysis of the cost drivers in the process.
- Relation of cost drivers to value to the customer.
- More detailed planning, specification and integration of the activities of each construction worker.

The potential disadvantages of the arrangement could be problems in co-ordination and difficulties though a middle-level organisation keeping the lead contractor from the subcontractors who are doing the work, and taking a margin for doing so. Ideally each of the services and fitout contractor should directly employ labour for their package of work. However, if they were not able to do so, because of fluctuations of workload, there must be incentives for the middle level contractors, in a three-level structure, to innovate and add value by their co-ordination.

Examples of construction process reengineering

This one of the three examples given in the T40 research report.

The three electricians

Under the current arrangements it is possible for three electricians, employed by different subcontractors, to be working on adjacent ladders in the suspended ceiling space. One is installing the lights, one installing the smoke detectors and one the air conditioning. They could each do 30 minutes work and then drive to the next project which, by chance, happens to include electrical tasks, and so on and so on. A much more efficient approach would be for each electrician to do the full electrical work on each project.

An extension of this logic is illustrated in Tables 5.4 and 5.5 which

Table 5.4 Traditional arrangements of subcontractors in the tenant space.

Activity	Current activities	By whom	Visit to area
1	Air conditioning ducting	A	1
2	Hydraulic rough-in	B	2
3	Sprinkler pipes	B	2
4	Steel stud partition frames to u/s slab	C	3
5	Electric cable trays	D	4
6	Smoke detectors rough-in	B	5
7	Electrical rough-in	D	6
8	Computer LAN rough-in	E	7
9	Telephone rough-in	F	8
10	Ceiling grids	G	9
11	Tiles to sprinkler heads, AC regulators and lights	G	9
12	AC fitout	G	9
13	Sprinkler fitout	G	9
14	Plaster board partitions	G	9
15	Aluminium partition frames	H	10
16	Glazing partitions	H	10
17	Joinery	I	11
18	Hang doors	I	11
19	First coat of paint to partitions	J	12
20	Skirting ducts	K	13
21	Carpet and vinyl	L	14
22	LAN fitout	E	15
23	Telephone fitout	F	16
24	Electrical fitout	D	17
25	Smoke detector fitout	B	18
26	Ceiling tiles	G	19
27	Second coat of paint to partitions	J	20
28	Work stations	K	21
29	Testing and commissioning power and lights	D	22
30	Testing and commissioning air conditioning	A	23
31	Testing and commissioning hydraulics and sprinklers	B	24
32	Testing and commissioning smoke detectors	B	25

Table 5.5 Proposed T40 arrangements of package subcontractors in tenant space.

Activity	T40 proposed activity	By whom	Visit to area
1	AC ducting Hydraulic and sprinkler rough-in Smoke detectors Cable trays Telephone rough-in Electrical rough-in	A	1
2	Computer LAN rough-in	B	2
3	Ceiling grid Tiles to sprinkler heads, AC registers, lights Stud partition Plasterboard Aluminium glazing frames Glazing Joinery Hang doors First paint Skirting ducts	C	3
4	LAN fitout	B	4
5	Mechanical and electrical fitout	A	5
6	Telephone fitout	D	6
7	Ceiling tiles, final coat to partitions	C	7
8	Install work stations	E	8
9	Testing and commissioning services	A	9

itemise the traditional arrangements of subcontractors in the tenant space and the proposed T40 arrangements of package subcontractors in the tenant space.

Workforce empowerment

There are strong moves in many sectors of manufacturing, and in some sectors of construction, to redefine the traditional role of the foreman from that of planner and supervisor to that of coach, thus encouraging workers to be responsible for their own work, the planning of that work, the purchase and delivery of material for that work, and related activities.

Specific aspects of workforce empowerment are workers who:

- are the eyes and ears of the foreman in identifying and correcting safety breaches and QA issues
- order their own equipment and materials
- are represented on workplace committees responsible for resolving any potential disputes on site
- are responsible for their own development to an agreed structure.

Teaming with the workforce

What is sought in a T40 environment is to achieve a fundamental cultural change which allows both the management and the general workforce to understand and respect each individual's or collective group's needs and subsequent goals.

The T40 solution embraces the concepts of teaming by:

- Encouraging participating companies to develop a company enterprise agreement, either as a certified agreement or a formal commitment/understanding, in relation to needs, goals and rewards.
- Establishing objectives and goals of not only the commercial stakeholders but participating workforce for the T40 project.
- Communicating this to all participants in the T40 project. They will often identify or relate to those that are common and can be identified in relation to their enterprise.
- Establishing methods of assessing the agreed objectives and goals that are being achieved.
- Encouraging workplace consultative committees and toolbox meetings to ensure continuing communication on what is or is not being achieved.

Codetermination

An extension of teaming is codetermination. A basic principle of codetermination is an absolute commitment to resolving industrial issues during the life of the project without resorting to lost time.

People innovation

An empowered group of people can be expected to make significant contributions to innovation on the project. Part of empowering people is to release these energies and thus encourage them to see the project as their own, so that they can contribute to the ideas bank on the project. Such arrangements could be the basis of a project enterprise agreement or an addendum to the company enterprise agreements.

Partnering with local government

The general theme of the T40 proposals is to partner with local government to satisfy both the needs of local government and the applicant. Some aspects of this proposal can be implemented immediately whereas some aspects will require longer-term changes.

Longer-term planning change

A longer-term planning approval solution is for the building use and envelope to be specified as part of the zoning and, provided the proposed project meets the preapproved guidelines, the planning approval could then be provided by administrative decision and action.

Tendering

The central assumption of the T40 process is that the full solutions team, including the general contractor and specialist contractors, are fully engaged in addressing the customer needs from the outset. This significant process requirement is quite different to lump-sum tendering and even the current form of document and construct, of developing a basic design by external consultants and then calling tenders based on this outline design and performance requirements. Both of these current processes exclude the solutions team from directly addressing the customer's need.

The proposed process eliminates the waste of current tendering, with success rates for contractor of approximately one in ten. What is needed is a system that selects the construction contractor on the basis of:

- Past record on time performance by reference to an industry time database.
- Agreed cost based on an adequately developed design.
- Third-party endorsement of cost as being close to the best price.
- Agree time which is significantly better than the industry average, for a building of the type (e.g. 40% better time performance, which corresponds to a 25% reduction in costs).
- Public accountability.

These points are illustrated in Table 5.6 on the next page.

Conclusion

In this chapter, we have traced the origins of reengineering, we have examined its goals and detailed the tactics which have been adopted by successful practitioners of reengineering in the business community at large. We have also included a detailed description of the T40 project, which, as a construction industry research study in process reengineering, exemplifies the key characteristics of the reengineering approach.

We have contrasted the directness of the North American approach to reengineering with the more restrained European view. Our own view of reengineering is one of cautious optimism. At present the very few examples of the successful application of reengineering in the construction industry prevent us from being more sanguine about its future. Having said that, we do note however that there is a groundswell in the industry towards the adoption of reengineering principles with the Commonwealth of Australia's Science and Industry Research Organisation (CSIRO), Division of Building, Construction and Engineering expressing a firm commitment to the reengineering of the construction process. To quote:

> 'CSIRO is committed to construction process re-engineering, recognising that it can significantly strengthen the Australian construction industry against overseas competition, both in the home and overseas market[1].'

The Egan report *Rethinking Construction*[28] has overtones of reengineering in its commitment to annual reductions of 10% in

Table 5.6 Differences between tradition, document and construct and the T40 solution.

	Fully document lump sum	Document and construct	T40 process
Contractor involvement in determining customer needs	No	No	Yes
Subcontracting involvement in determining customer needs	No	No	Yes
Single point accountability	Yes	Yes	Yes
Number of subcontractors and suppliers	50–100	50–100	Tiered with 8–10 in first layer
Incentives and penalties for subcontractors	No	No	Yes
Identification of subcontractors with owner	No	No	Yes
Clear stages of project approvals	Sometimes	Sometimes	Yes
Real-time design	No	Sometimes	Yes
Variations in process	Usually many	Medium to few	Nil
Documentation process	Multi-origins	Some co-ordination	Staged, packaged and complete
Project communications	Phone & fax	Phone & fax	Direct
Use of specialist resources	Separate	Separate	Shared
Supervision	Separate	Separate	Shared
Automatic invoicing	Seldom	Seldom	Yes
Tendering	Costs of 6–8% and time of up to 3 months	Costs of 3% and time of up to 3 months	Virtually zero costs
Local authority approvals	6–12 months	6–12 months	0 months
Predictable outcome	No	Better	Yes
Overall work to complete	W	90%W	75%W
Time outcome	T	90%T	60%T

construction cost and construction time and a reduction of 20% per annum of defects in projects. This represents a total performance improvement of 30% (which is not dissimilar in objective to the Australian T40 project). One initiative arising from the Egan report was the commissioning of research by the Engineering and Physical Sciences Research Council (EPSRC) which was directed at developing a Generic Design and Construction Process Protocol (GDCPP) by reengineering the design and construction process. This work is being undertaken at the University of Salford by Kagioglou, Cooper and Aouad[29] and is ongoing. (The Process Protocol website at Salford University[30] is regularly updated and is a useful source of information on the reengineering movement in the UK.)

The Centre for Civil and Construction Engineering of the University of Manchester Institute of Science and Technology (UMIST) is currently undertaking development work in performance based business process reengineering (now entitled reengineering construction) as part of an International Council of Research and Innovation in Building and Construction (CIB)[31]. The aim of this work is to produce a strategy to increase the uptake of construction reengineering.

Reengineering has the power to change the very structure and culture of the construction industry. Time will tell whether or not the industry will rise to the challenges posed by the reengineering approach. Interestingly the CIB announcement of its reengineering initiative also poses the question that perhaps reengineering construction is already a dated concept. We rest our case.

References

1. Love, P. & Mohamed, S. (1995) Construction process reengineering. *Building Economist*, December 8–11.
2. Hammer, M. & Champy, J. (1993) *Reengineering the Corporation: A Manifesto for Business Revolution*. Nicholas Brealey, London.
3. Morris, D. & Brandon, J. (1993) *Re-engineering your Business*. McGraw-Hill, New York.
4. Coulson-Thomas, C.J. (ed.) (1994) *Business Process Reengineering: Myth and Reality*. Kogan Page, London.

5. Hammer, M. (1990) Re-engineering work: Don't automate, obliterate. *Harvard Business Review*, July/August, 104–12.
6. Rigby, D. (1993) The secret history of process engineering. *Planning Review*, March/April, 24–7.
7. Whitney, J.O. (1987) *Taking Charge: Management Guide to Troubled Companies and Turnarounds*. Dow Jones-Irwin, Illinois.
8. Greengard, S. (1993) Re-engineering: Out of the rubble. *Personnel Journal*, December.
9. Mohamed, S. and Tucker, S.N. (1996) Construction process engineering: potential for time and cost savings. *International Journal of Project Management* (special issue on Business Process Re-engineering).
10. Construction Industry Development Agency. (1994) *Two Steps Forward and One Step Back: Management Practices in the Australian Construction Industry*. CIDA, Commonwealth of Australia Publications, Sydney, NSW, February.
11. MacDonald, J. (1995) *Understanding Business Process Reengineering*. Hodder & Stoughton, London.
12. Obeng, E. & Crainer, S. (1996) *Making Re-engineering Happen: What's Wrong with the Organisation Anyway?* Pitman Publishing, London.
13. DuBrin, A.J. (1996) *Reengineering Survival Guide: Managing and Succeeding in the Changing Workplace*. Thomson Executive Press, Ohio.
14. Armstrong, J.S. (1985) *Long-range Forecasting: from Crystal Ball to Computer*, 2nd rev. edn. Wiley, New York.
15. COBRA (1994) *Business Restructuring and Teleworking: Issues, Considerations and Approaches*. Methodology Manual for the Commission of the European Communities, London.
16. Coulson-Thomas, C.J. (1994) Implementing re-engineering. In: *Business Process Reengineering: Myth and Reality* (ed. C.J. Coulson-Thomas) pp. 105–26. Kogan Page, London.
17. Mohamed, S. & Yates, G. (1995) Re-engineering approach to construction: a case study, *Fifth East-Asia Pacific Conference on Structural Engineering and Construction – Building for the 21st Century*. 25–27 July, **1**, 775–80, Gold Coast, Queensland.
18. Petrozzo, D.P. & Stepper, J.C. (1994) *Successful Reengineering*. Van Nostrand Reinhold, New York.
19. Fitzgerald, B. & Murphy, C. (1994) The practical application of a

methodology for business process re-engineering. In: *Business Process Reengineering: Myth and Reality* (ed. C.J. Coulson-Thomas) pp. 166–73, Kogan Page, London.

20. Betts, M. & Wood-Harper, T. (1994) Re-engineering construction: a new management research agenda. *Construction Management and Economics*, **12**, 551–6.
21. Holtham, C. (1994) Business process re-engineering: contrasting what it is with what it is not. In: *Business Process Reengineering: Myth and Reality* (ed. C.J. Coulson-Thomas) pp. 166–73, Kogan Page, London.
22. Hazlehurst, G. (1995) Re-engineering the processes of construction engineering and design: islands of automation. *COBRA 95 RICS Construction and Building Research Conference*, 8–9 September, **2**, 207–15, Heriot-Watt University, Edinburgh.
23. Venkatraman, N. (1991) IT-induced business reconfiguration. In: *The Corporation of the 1990s* (ed. Scott Morton) pp. 122–58, Oxford University Press, New York.
24. Newton, P.W. & Sharpe, R. (1994) Teleconstruction: an emerging opportunity. *National Construction and Management Conference*, February, 447–59, Sydney.
25. Mohamed, S. Simulation of construction site operations: a re-engineering perspective. Submitted to *Engineering, Construction and Architectural Management Journal.*
26. Talwar, R. (1994) Re-engineering: a wonder drug for the 90s? In: *Business Process Reengineering: Myth and Reality* (ed. C.J. Coulson-Thomas) pp. 40–59, Kogan Page, London.
27. Ireland, V. (1994) *T40 Process Re-engineering in Construction.* Research Report, Fletcher Construction Australia Ltd, May.
28. Egan, J. (1998) *Rethinking Construction*, DETR, London.
29. Kagioglou, M., Cooper, R. & Aouad, G. Re-engineering the UK construction industry: the process protocol. *Second International Conference on Construction Process Reengineering*, 12–13 July 1999, Sydney, Australia, 425–36.
30. Process Protocol Salford University website (2002) http://www.Salford.ac.uk/gdcpp.
31. CIB website (2002) http://www.cibworld.nl/pages/begin/Pro4.html.

Chapter 6
Total quality management

Introduction

In an article in *The Times*[1] newspaper, a houseowner claimed that after moving into a newly purchased house it took the builder, a national contractor, 103 days to finish the kitchen. In addition, in the short time they had lived in the house they found 112 faults. The couple paid £82,000 for the property but despite this the house was not ready on the day of completion of the purchase and the painters were still in the house on the day they moved in. The large number of faults that appeared meant that for five months workmen from the contractor visited the house three or four times a week. The house owner admitted in the article that the contractor had done all they could to put the defects right. Nevertheless the problems kept appearing.

Naturally the above is a fairly extreme example and in isolation probably not a major problem. However what was really remarkable about the article was the advice of the National House Builders Federation to potential purchasers of new properties. This included:

- ❏ Buy through a solicitor or surveyor.
- ❏ Take care with your contract as recourse will depend on its terms and conditions.
- ❏ As well as thoroughly reading the contract buyers should, with the aid of a solicitor, help to write the contract.

The National House Building Council, which is the standard

setting and regulatory body of house building also offered their advice:

- ❑ Check the builder is registered with the NHBC.
- ❑ Talk to previous customers to check the builder's standards.
- ❑ Check the site is clean and well-managed.
- ❑ Turn up on site at an unexpected time.
- ❑ A good sign is a fluttering flag which says that the site manager has won an award under the NHBC pride in the job scheme.
- ❑ Outside the house look for problems with brickwork, roofing, paintwork, pipework, pipes and drains. A further list is included for the inside of the property.
- ❑ When there are defects with a property the first port of call is the builder provided he is still in business.

This advice is, in the authors' view, indicative of the endemic problem of poor quality in the construction industry. It is remarkable that these two national bodies, supposedly there to protect the interests of house purchasers, really expect that the average member of the public is able to distinguish a clean and well-managed construction site from a dirty and poorly managed one. It is our guess that to the average member of the public all construction sites are dirty. Furthermore why does the potential purchaser need to buy a house through a surveyor. The national contractor mentioned in the article has been building and selling houses for decades. Why can't they do a decent job of selling homes themselves?

Suppose for a moment that the house in question were a car. Imagine if you bought a car for tens of thousands of pounds. Would the manufacturer allow you to drive it away when the paint was still wet? When you buy a new car, do you need a solicitor to help you interpret and redraft the contract? Do you need to turn up unexpectedly at the factory to check it is clean and well-managed? Would you be able to distinguish a clean car factory from a dirty one? And if it turned out that the car had 112 defects, what would you do? Would you allow the manufacturer to come to your house three or four times a week for five months to fix it? Highly unlikely. More likely you would eventually demand a new car and in all probability the manufacturer, realising the problems were unacceptable, would meet your request.

The problem that the construction industry has is one of poor quality culture. This is reflected by the advice of the two national bodies. The whole industry appears to start from the standpoint that the customer has to look after him or herself and that it is not necessarily the job of the contractor (or the consultants) to do that.

There can be no doubt that in terms of quality the construction industry has improved enormously in recent years but this improvement appears only to be in regard to certain aspects of the construction process. Sites may be better managed and houses may be better designed than they were ten years ago but final delivery is still not good enough. This is because the industry is not truly customer focused. The industry does not see quality as a whole issue driven by satisfying customer need but as a series of procedures dealing with design, materials or site safety. How *can* a national building contractor with any degree of customer focus allow a buyer spending £82,000 to move into a house that is in the process of being painted?

It is not only the construction industry that is guilty of this lack of customer focus. Manufacturing organisations as well, under competition from Japanese products, have paid the price of poor quality in terms of lost market share. Many of these organisations have now recognised that true quality comes not from a series of checks on products and standards but from a holistic approach to quality that is customer focused. Organisations in the service sector have been particularly successful. National supermarkets are a good example of a holistic approach to quality, with some stores now even offering Sunday school for the children while parents shop.

This then is the essence of total quality management (TQM). It is a customer-focused total approach to quality that involves all aspects of, and all people involved in, the organisation.

Definition of TQM

Of all of the techniques outlined in this book TQM is the broadest and most wide-ranging. In some ways it is an umbrella under which all other concepts are encompassed (this will be discussed later). It is possibly for this reason that there is no single accepted definition of TQM. Rampsey and Roberts[2] for example define it as:

> 'A people focused management system that aims at continual increase in customer satisfaction at continually lower real cost. TQM is a total system approach (not a separate area or program), and an integral part of high level strategy. It works horizontally across function and department, involving all employees, top to bottom, and extends backwards and forwards to include the supply chain and the customer chain.'

Compare this with the following definition[3]:

> 'TQM is the integration of all functions and processes within an organisation in order to achieve continuous improvement of the quality of goods and services. The goal is customer satisfaction.'

These definitions are in fact quite similar, in that both encompass the fundamental principles of TQM. First, TQM must be a total approach to quality. Whereas in the past quality was concerned with parts of the organisation, such as the final product or customer relations, TQM is concerned with the whole system as an integrated unit. Second, TQM is ongoing. Whereas in the past quality was viewed as a system which could be put in place to improve certain sections of the product or organisation, TQM is a continuous process. The view is now taken that however good a system has become, it can always be improved further. Finally the goal of TQM is customer satisfaction. In the past quality systems have been aimed at improving products and not at customer satisfaction. These are not necessarily the same thing, since it is possible to improve a product without realising that it is not in fact what the customer wants.

Figure 6.1 summarises the three drivers of TQM: integration, customer focus and continuous improvement.

Before examining these three drivers in detail it is worth considering some underlying ideas that support them. First is an examination of what constitutes quality. Second is the historical development of TQM and third is the idea of organisational culture.

What is quality?

As in the case of TQM there are many different definitions of quality[3].

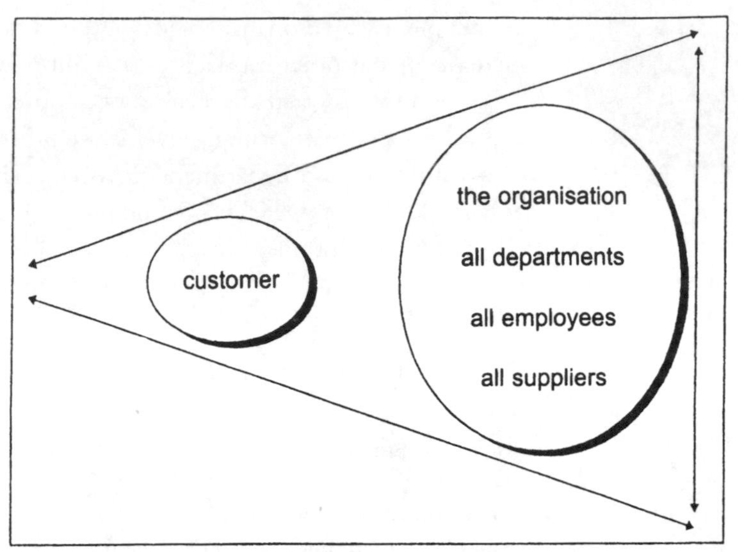

Fig. 6.1 Three drivers of TQM: integration, customer focus and continuous improvement.

Manufacturing-based definitions view quality as the ability to conform to requirements or specification. This measure of quality is objective, in that it is based purely on the ability of the product or service to meet a predefined specification or standard. We might, for example, measure if an electric fire produces the correct output of heat or whether a percentage of construction projects were completed on time. The problem with this type of quality definition is that there is no indication that what is measured is in fact what the customer wants. It is an inward-looking measure of quality that could not be defined as a total quality approach.

Product-based definitions of quality are also objective in that they are based on a measure of a specific characteristic of a product such as, for example, durability or maintenance. A UPVC window could be said to be better quality than a timber window because it lasts longer and needs less maintenance.

User-based quality definitions, on the other hand, are subjective and evaluate quality based on the extent to which a product satisfies the user. This concept of user satisfaction is closely allied with the idea of function analysis examined in the chapter on value management. User quality can be described as that quality which satisfies the user. In terms of construction this may mean that an item of lower product quality may have higher user quality. In a warehouse building for example the UPVC window mentioned above may have higher product quality. However if the building is low specification, only designed to last 20 years, then timber windows may well have higher user quality because they better suit the needs of the user to provide windows that will last 20 years.

Value-based definitions of quality tend to include one of the measures of quality stated above but in the context of cost. A value-based examination of the quality of UPVC and timber windows would therefore recognise that the UPVC products are more durable and easier to maintain but this would be viewed in the context of their higher cost. This 'best buy' approach to quality is the one often used by consumer magazines.

So for the modern construction organisation, which of the above definitions of quality should be adopted? The answer is that all of them have a role to play. A good construction project will conform to specification and satisfy the user with given levels of quality of the attributes required at the desired price. This may sound rather abstract but these diverse concepts of quality can be translated into a more objective list of principal quality dimensions[4].

Performance

This is the primary reason for having the project along with the main characteristics it must have. In terms of a hospital this may therefore be the provision of wards, waiting rooms and operating theatres.

Reliability

This asks if the building will operate for a reasonable period of time without failure.

Conformance

This is the degree to which specification is met.

Durability

This is the length of time a building lasts before it needs to be replaced.

Serviceability

This is the service given after the building is completed, particularly with regard to repair.

Aesthetics

This is how the building looks and feels.

Perceived quality

This is the subjective judgement of quality that results from image. Table 6.1 below examines these quality dimensions further. How well does an average building measure up?

Table 6.1 Principal quality dimensions.

Quality dimension	Building performance
Performance	Do the majority of buildings achieve their main purpose?
Reliability	Are they reliable?
Conformance	Do they conform to the specification?
Durability	Do they last a longer or shorter period than is required?
Serviceability	Are they repaired quickly and with a quality service?
Aesthetics	Are they aesthetically pleasing internally and externally?
Perceived quality	Does the user and client feel it is a quality building?

Quality therefore is not an easy word to define. It exists at different levels, ranging from the degree to which components of the building meet specification, to the degree that the whole building satisfies the customer. It may be judged either in isolation or relative to some other objective measure such as cost. The quality of a building can be evaluated based on a list of seven principal dimensions. The way to achieve high levels of these principal dimensions is through management of all the processes that deliver them. This is the nature of TQM.

Historical development of TQM

The idea of quality is not new and has its origins in inspection-based systems used in manufacturing industries[5]. In order to reduce the number of faulty goods passed on to the customer, products were inspected during the manufacturing process. The products under inspection were compared with a standard and any faulty goods not reaching the standard were weeded out and either scrapped, or repaired and sold as seconds. These types of inspection-based quality systems were found to have several disadvantages:

- Where an inspection did not reveal the faulty item, the problem was passed on to the final user or customer.
- Inspection-based systems are expensive because they are based on rectifying faults.
- Inspection-based systems remove responsibility from the workers and place it on to the inspectors.
- Inspection-based systems give no indication of why a product is defective.

For these reasons and also because products were becoming more complex, inspection-based systems were replaced by systems of quality control based on statistical sampling. One of the gurus of these systems was Deming[6]. The main focus of Deming's work was improvement of the product by reduction in the amount of variation in design and manufacturing. To him variation was the chief cause of poor quality. He believed that variation came from two sources: common causes and special causes. Common causes were those that came about as a result of problems in the production process, whereas special causes were a result of a specific individual or batch of material. In construction terms a common cause might include the sealing of baths against glazed tiles, since there is a fault with the actual design itself and not the material or workmanship involved. A special cause, on the other hand, might include a bricklayer mixing the mortar in too weak a mix.

In order to achieve improvement in quality through reduced variation Deming outlined a 14-point system of management. These points focus on the process, in that Deming believed it is systems and not workers which are the cause of variation. His points were:

(1) Create and publish the aims of the company
(2) Learn the new philosophy of quality
(3) Cease dependence on mass inspection
(4) Do not award business based purely on price
(5) Constantly improve the system
(6) Institute training
(7) Institute leadership
(8) Drive out fear and create trust
(9) Break down barriers between departments
(10) Eliminate slogans and targets
(11) Eliminate numerical quotas
(12) Remove barriers to pride in workmanship
(13) Institute self-improvement and a programme of training and retraining
(14) Take action to accomplish the change.

Deming believed that once a quality system was set in motion it resulted in a quality chain reaction (figure 6.2). That is, as quality improves costs decrease, as do errors and delays. This causes an increase in productivity and an increase in market share brought about by better quality at lower price. This means the company is more competitive and provides more employment[4].

Along with Deming, Juran, another American quality consultant, introduced quality control techniques to the Japanese[4]. As the con-

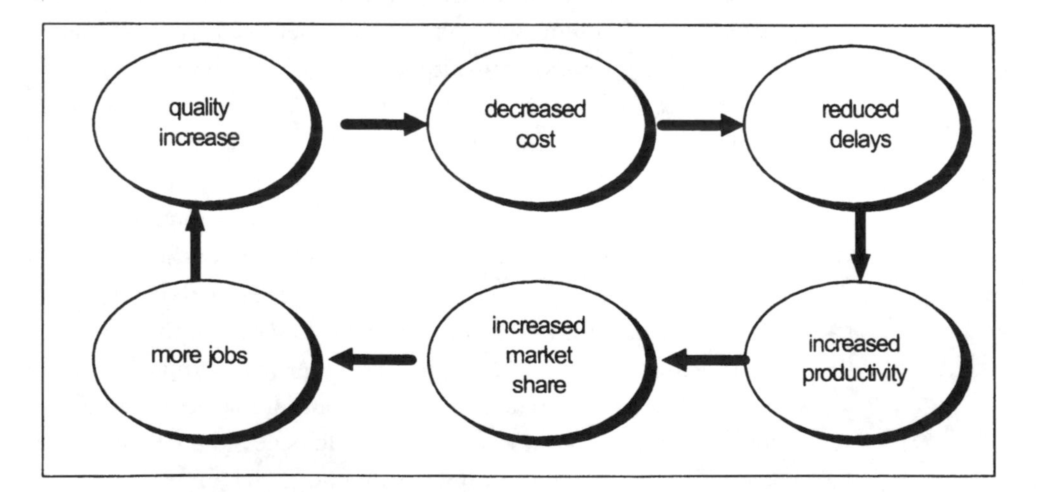

Fig. 6.2 Quality chain[6].

sumption of manufactured goods continued to rise over the next three decades Japanese products came to dominate Western markets. This was because Japanese goods, largely because of their superior quality systems, were of higher quality than their Western counterparts. Whereas quality management had developed as a subject in its own right in Japan, Western quality methods had remained unchanged since their introduction in the 1950s.

There can be little doubt that this gap in quality between American and Japanese goods was one of the drivers of the quality revolution that is taking place in the US and Europe. The word revolution is not used lightly, as recent years have indeed witnessed a fundamental change in the way quality is viewed. Whereas prior to the 1980s quality was internally focused, it is now customer or externally driven. The concept of quality is now strongly related to the idea that an organisation is a series of processes and that for each of these processes there are customers. These customers may be internal to the organisation but by satisfying them the end product will be improved and the end user satisfied. In addition, and as seen from the definitions given above, TQM is now viewed as a much broader philosophy than the traditional techniques of quality control. It now encompasses all of the organisation from senior management right through to customers and suppliers. Finally, whereas the older techniques of quality control relied on inspection, TQM relies on prevention[6].

It would be wrong to assume that Deming was the sole guru of quality; other practitioners have disagreed with his approach. Juran's idea of quality, for example, was not as wide-ranging as Deming's and rather than attempting radical change, he sought to improve quality by working within the existing system of the organisation. Crosby's quality philosophy was different again, placing greater emphasis on behavioural aspects than on the statistical analysis used by Deming[4].

The ideas put forward in this chapter therefore are not based solely on the ideas of Deming but on a combination of ideas that are understandable in the context of the construction industry. However this work does not suggest a rigid procedure for TQM because no such thing exists. TQM is a philosophy not a technique. A philosophy implies a way of thinking and in this context TQM offers its biggest challenge. The way that people think is determined by their culture and in order to instil TQM in the company it is necessary for the organisation to undergo a cultural change. This idea of a cultural shift is closely allied to a modern idea in construction management: that of the paradigm shift.

The need for a paradigm shift

When an industry is developing it is working within the boundaries of its existing system. Any changes that are made, or any developments that take place, do so within those boundaries. These boundaries are useful in that they illustrate the accepted opinion of the discipline and the rules within which it operates. However at some point in history these boundaries become too restrictive and it is necessary to shift the boundaries in order to either explain the changes that are taking place or to facilitate the development that is needed. This shifting of the boundaries is known as the paradigm shift. Many writers on TQM and on the other concepts in this book argue that the construction industry is at the point where the paradigm shift is taking place and that a new paradigm is emerging to replace the old one. It is recognised that the existing paradigm is no longer adequate to meet the needs of a competitive and global construction industry and that it must change. Table 6.2 illustrates this idea of the new and old paradigm[7]. At the beginning of this chapter customer focus was outlined as one of the key drivers of TQM and the table shows paradigms of customer focus broken down into the topics of quality, measurement, position, key stakeholding and product design.

An organisation operating under the old paradigm would therefore be a follower, building products it knows will sell. Its main

Table 6.2 The new and old paradigms.

Topics	Old paradigm	New paradigm
Quality	Meeting specification	Customer value
Measurement	Internal measure of efficiency	Linked to customer value
Positioning	Competition	Customer segments
Key stakeholder	Stockholder	Customer
Product design	Internal sell what we build	External build what the customer wants

priority would be profit for its shareholders and senior management and it would judge its success against the success of other organisations. It would measure efficiency against predetermined figures with little investigation into what causes the figures and it would control quality through inspection. Finally it would regard quality, in the main, as the meeting of specification. An organisation operating within the new paradigm, on the other hand, is one that finds out what the customer wants and builds products that meet those needs. Quality and measures of efficiency are all linked to customer value, and positioning in the market is geared towards market share. Finally, short-term profits are not the aim as it is the customers, not the shareholders, who are seen as the key stakeholders of the organisation.

In the introduction to this chapter it was outlined that TQM had three drivers: integration, customer focus and continuous improvement. Table 6.2 deals only with one aspect of the new paradigm, namely customer focus. In addition new paradigms are emerging in terms of integration and continuous improvement. In terms of integration, barriers between organisational functions are breaking down and human and behavioural aspects of the organisation are more becoming important. Tall hierarchical structures of the past are being replaced with flatter structures with an emphasis on teamwork. With regard to continuous improvement the move is away from top-down management operating in a short-term perspective, towards more open systems of management concentrating on long-term plans that are customer driven[7].

A change in the culture of the construction industry

Paradigms refer to the ways of thinking and acting that are characteristic of a body of knowledge. Culture refers to the social structures that underpin those thoughts and actions. TQM requires a paradigm shift in order to be successfully implemented. For that to happen there needs to be a cultural shift in the organisations that make up the construction industry. This cultural shift that it requires can be summarised in Table 6.3[8].

Organisational culture however is a reflection of social culture and

Table 6.3 The cultural shift required for TQM.

From	To
Meeting specification	Continuous improvement
Complete on time	Satisfy customer
Focus on final product	Focus on process
Short-term view	Long-term view
Inspection-based quality	Prevention-based
People as cost burdens	People as assets
Minimum cost suppliers	Quality suppliers
Compartmentalised organisation	Integration
Top-down management	Employee participation

this means that it is difficult to change, since it is vested in the rules that hold society together. However it is not impossible to introduce new organisational culture into the work environment. Evidence of this can be seen in the success of Japanese companies operating in the UK who have successfully planted their organisational culture, albeit with variation, into a social system wholly different from their own[9].

A Japanese company opening a new plant with its own organisational culture however may be easier than a large British construction company changing from one organisational culture to another. However by realising that a change in organisational culture *is* possible and that it is the only way to achieve TQM can be a greater motivator for change. In addition there are ways in which organisational culture can be changed. Change can be achieved through[7]:

- ❑ Management roles
- ❑ Training and developments
- ❑ Processes and systems
- ❑ Employee involvement.

Looking at one of these aspects, management roles, in more detail, the traditional view of management's role was one of

- ❑ Planning
- ❑ Organising
- ❑ Commanding
- ❑ Co-ordinating
- ❑ Controlling.

However this now needs to include:

- ❑ Figurehead
- ❑ Leader
- ❑ Liaison
- ❑ Monitor
- ❑ Disseminator
- ❑ Spokesperson
- ❑ Entrepreneur
- ❑ Disturbance handler
- ❑ Resource allocator
- ❑ Negotiator

This is a broader role, which is integrated into the organisation, is more people-focused and is a much more realistic assessment of the modern management role. Organisations that wish to adopt TQM must therefore recognise that not only is a cultural change required but also that the changing of culture is not the impossible task it is often assumed to be. A proactive approach to cultural change is outlined in table 6.4[10] which also shows an example of how such a strategy might operate in the construction industry.

Table 6.4 A proactive approach to cultural change.

Strategy	Example
Examine the present	Poor subcontractor relationships
Identify the desired value system	Good subcontractor relationships
Develop policies to embody value system	Use a smaller group of subcontractors continuously Set up internal mediation procedures to deal with subcontractors' grievances Do not hire subcontractors without first inspecting work that they have done Adopt a fair working charter
The total quality strategy	Compile group of subcontractors Devise procedures for mediation Draw up fair work charter
Implement total quality management	Communicate new policy to subcontractors Train employees in new methods and policy

The first section of this chapter examined some of the ideas that underpin total quality management, not least the concept of quality and the need for a cultural shift within the construction industry. The next section concentrates on the three main drivers of TQM: namely customer focus, integration and continuous improvement.

Customer focus

The idea of customer focus is important because it gives managers a means of providing products and services that meet customer need. The problem with construction is that it is often difficult to define who the customer is. From the point of view of TQM there are two types of customer: the end user and the internal customer. However in construction even the end user may not be easy to define, since there are often many end users each with conflicting needs. The technique of value management however is geared towards optimising the needs of all the end users of the project. Although this is a new idea, the authors recommend that value management be incorporated into a TQM programme as, in our view, this technique above all others is capable of the complex process of rationalising the conflicting interests that are involved in the construction project.

The idea of the internal customer is one that has been examined before, particularly in the chapter on benchmarking. The idea of the internal customer relates strongly to the idea of the business process. TQM, like benchmarking and reengineering, is process-oriented. As such the internal customer is the customer of the process or the next person down the chain of production. How to improve the service offered to the internal customer is examined in the next section on integration.

Integration

Traditionally organisations have been separated into departments that carry out a set function, such as estimating. They have also, within these functions, been arranged hierarchically. The need to divide companies this way is accepted practice despite the fact that it leads to some of the following problems[10]:

- Empire building
- Inbreeding
- Diversification of values
- Competition between departments
- Conflicting priorities
- Different expectations
- Rigid structures.

An organisation that is integrated, on the other hand, has a single objective and a common culture. Communications are improved and there is respect for the individual and not the department they work in. One of the main drivers of TQM aims is to break down the compartmentalisation of organisations and move towards more integrated structures. This can be achieved in several ways.

Process management

As explained above an organisation has internal customers who are the customers of the business process. This concept of the business process is central to the idea of TQM, since it is believed that if all internal customers are satisfied, the product or service offered to external customers or end users will also be improved. On a construction project there are usually several organisations involved and this is often a cause of major managerial problems. However the idea of the business process cuts across these boundaries as it is assumed that all the people, in all the companies, are customers. If the main contractor viewed the subcontractor as a customer (and vice versa) it is assumed that the construction process would run much more smoothly and that a better service and project would be provided to the client and end users. Despite the commonly held but contrary view of the construction industry this is really no different from manufacturing, where different companies manufacture components that make up the final product. Manufacturing industries however appear to accept to a much greater extent than the construction industry, that the overall quality of the final product is only a reflection of the quality of all the materials and workmanship that have gone into producing it regardless of the company.

In terms of TQM therefore a customer is defined as anyone who has the benefit of the work, activity or actions of another. Customers can be categorised as:

- Internal customers who are customers of processes and are within the organisation. (Sales department)
- External customers who are customers of processes and are outside the organisation. (Subcontractors)
- End users who receive, and pay for, the final product. (The client)

It is assumed that satisfaction of internal and external customers through improvement of processes will lead to a higher standard of final product. The idea behind this is that it is not output which should be quality controlled but processes. This is a reversal of the traditional patterns of Western management which have tended to examine output and be highly results-oriented. Western systems are generally top-down and tend to be designed by management who implement them and wait for results, against which they measure success or failure. This can be described as unitary thinking: that is, decisions based totally on quantitative measures of output. The quantitative measures generally have some financial point of reference.

The effects of unitary thinking[10]

Unitary thinking is geared towards numbers but this is a very poor way of examining the organisation and tends to lead to:

Stagnation

If an organisation constantly uses the same targets to measure its success it will undoubtedly stagnate. The environment within which an organisation exists is constantly changing. It makes no sense therefore for the targets against which success is measured to remain constant.

Undermining of quality

Quality cannot be measured purely quantitatively. Reliance on numerical measures of success therefore will undoubtedly lead to the focus moving away from quality and on to achievement of the numerical target.

Shortening timescale

Where a numerical target has to be met there is a tendency to focus on the timescale within which it exists. Measuring success against quarterly reports or interim valuations encourages staff to see projects (and the company at large) as a series of short steps or phases. This type of short-termism is detrimental to quality and the long-term health of the organisation.

Inhibition of investment

Concentration on short-term results and targets discourages long-term investment.

Compartmentalisation activity

This is a common problem with large organisations and occurs when employees identify with their own particular department or activity and not with the organisation as a whole. This leads to infighting and lack of commitment to the company as a whole.

Process thinking

The reverse of unitary thinking is process thinking which, rather than concentrating on outputs and numerical measures of it, concentrates on processes. Behind process thinking is the idea that the cause of success and failure lies in the systems that produce the results and ultimately in the processes that make up the systems. The benefits of process thinking include[10]:

Lengthen timescales

Process improvement by its nature is ongoing. As a result a company that is process-oriented has a longer-term perspective which tends to promote investment and long-term success.

Create a company culture

As outlined earlier, at the root of TQM is the need for a cultural shift within the organisation. Concentrating on the process can assist it to make this cultural shift as it diverts attention away from output, short-term results and compartmentalisation.

The inverted organisation

The inverted organisation can also act as a means of integrating the organisation. Rather than seeing the organisational structure with the manager at the top, the inverted organisation views those who deliver the service as being at the top, as they are from the front line.

Employee involvement

A further way that integration can be achieved in the organisation is through employee involvement. This idea of all company employees being part of the organisation is not simply a utopian ideal, there for its own sake. Its purpose, through empowering members of the organisation to make decisions and solve problems, is to act as a means of improving the system. Its purpose is to pull employees into the process and to encourage ownership of it. Employee involvement can be achieved in any of the following ways.

Quality circles

Quality circles bring together those employees involved with a particular process to identify, analyse and solve job-related problems that relate to that process. Quality circles are important because they encourage ownership, and as a result improvement, of the process.

Quality circles can be negative or positive[10]. Negative circles are introduced only where there is a particular problem with the process. Positive circles, on the other hand, are permanent groups. Positive quality circles are viewed as the better of the two since there is often more to be learned from examining the process when it is normal than when it is problematic.

Essentials of a quality circle

- ❑ People
- ❑ Skills
- ❑ Time
- ❑ Place
- ❑ Resource

Stages in a quality circle organisation

- ❑ Identification – the process that is either problematic or which is to become the subject of a positive quality circle is identified.
- ❑ Evaluation – any problems or potential problems with the process are identified. A distinction is made between problems that can be rectified and those that are outside the control of the company.
- ❑ Solution – an optimum solution for improving the process is arrived at.
- ❑ Presentation – the results of the quality circle and the proposed improved system are presented to senior management.
- ❑ Implementation – the improved process is implemented. However as process improvement is a continuous process the new process must be monitored and if necessary improved further.

Quality circles are particularly useful in solving problems related to direct labour[3] where the problems dealt with are of a practical nature. There have been problems with their use in more general areas where middle management can see them as a threat to their authority.

Cross-functional teams

Cross-functional teams also increase integration by operating laterally in the organisation as opposed to following the typical hierarchical structure. The basis of cross-functional teams is the co-ordination of organisational functions on the basis of processes that run through them.

Training and development

There cannot be employee involvement without training. TQM training will usually fall into three categories of reinforcement of the quality message, skill training relevant to a particular task and principles of TQM[3]. Critical to training and development is problem-solving. Problem-solving is a huge area of management and for that reason only a brief outline of the major problem-solving techniques is outlined below.

Brainstorming

Brainstorming was outlined briefly in the chapter on value management as a means of generating creative design solutions. Brainstorming can also be used as a means of problem-solving. Like its counterpart in value management it has certain rules. Before starting a brainstorming problem-solving session there should be no definition of the problem other than in totally neutral terms and suggestions of solutions should not be made. Time for suggested solutions to the problem should be limited and all answers should be noted, however outrageous they may seem. Once again unusual ideas should be encouraged as these often generate effective and workable solutions.

Pareto analysis

A Pareto analysis involves calculating both the costs and frequency of problems that occur and putting them in order of magnitude so that the highest cost problems are revealed and singled out for improvement.

Cause and effect diagrams

These were invented by Kaoru Ishikawa and are also referred to as fishbone diagrams or Ishikawa diagrams[10].

In the chapter on benchmarking it was outlined that a process has an input, a process and an output and that an organisation consists of hundreds, or possibly thousands, of inter-related processes. In the fishbone diagrams a cause is an input to the process. An effect, on the other hand, is the result of the process. When the process is working correctly, or is in control, the effect or output of the process is predictable. When the process is out of control the effect is not predictable. Ishikawa believed that the investigation of cause is most critical. He believed that although something may look like a cause, in reality it is an effect of a previous process. By tracing back along the fishbone this effect can be found. It can only be found however when it is known what the true effect should have been. For that reason it is equally important to study the process while it is in control as it is when it is out of control.

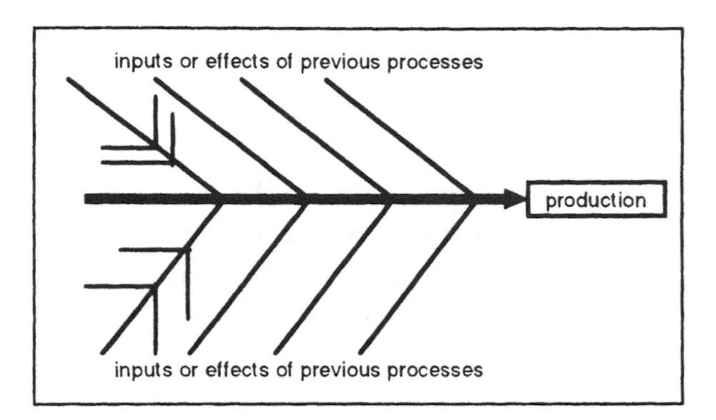

Fig. 6.3 Ishikawa diagram.

Performance appraisal

Performance appraisal is also a method of encouraging employee involvement. However, traditional methods of appraisal may not be suitable for a company on track to TQM. Performance appraisal should focus on the objectives of the company and not on the individual output.

The one-more theory

The one-more theory operates on the basis that every job in the organisation should be able to be undertaken by at least two people and that every person in the organisation should be able to do at least one other job in addition to their regular work.

The all-embracing nature of TQM

The all-embracing nature of TQM deems it necessary for companies to include both suppliers and customers in their TQM process. Of all the systems covered in this book it is probably TQM more than any other that falls into the category of a soft system. For that reason it is impossible to say what is a typical system of TQM. It is however possible to say what are the characteristics of a company *en route* to total quality[5].

- Vision
- Attention to customer
- Attack on failure
- Focus on prevention
- Zero functional barriers
- Monitoring of competition
- Senior management actions
- High levels of training.

Continuous improvement

The third driver of TQM is that it is a continuous process of improvement. Continuous improvement means that even where an organisation is profitable, with a high percentage of market share, it should still look for ways of improving. It cannot be iterated enough that TQM is not a system but a philosophy, an important part of which is the notion that no system is ever perfect and can always be improved.

Quality costs and the cost of quality

Many people reading this book may by now have concluded that their own organisation cannot, for reason of size or profit margins, afford to adopt TQM. However such opinions are often vested in a misunderstanding of what the actual costs of quality are. There are basically three ways of thinking about quality costs and these are outlined below.

Higher quality means higher cost

Increased quality cannot be obtained without an increase in cost and the benefits to the company from increased quality will not compensate for the additional cost involved.

The cost of improving quality is less than the resulting savings

This second view of quality costs is that the additional cost of quality is less than the money lost through the rebuilding or scrapping of work.

The right-first-time approach

This view of quality cost is much wider than that stated above. Quality costs are viewed as those incurred in excess of those that would have been incurred if the product were built right first time. Costs are not only direct but those resulting from lost customers, lost market share and other hidden costs. Costs are not measured as the costs of rebuilding or scrapping but the extra over cost that is incurred based on what the costs would have been if everything had been built the first time to zero defects. This is the view of most TQM practitioners.

These costs of quality measured against the zero defect can be classified into four categories:

- **Prevention** – These are costs which remove or prevent defects from occurring. Examples are quality planning and training.
- **Appraisal** – These are the costs incurred to identify poor quality products after they occur but before shipment to customers. Inspection costs are an example.
- **Internal failure** – These are costs incurred during the production process and would include scrapping and rebuilding costs.
- **External failure** – These are the costs of rejected or returned work and include the hidden costs of customer dissatisfaction such as loss of market share. Most TQM practitioners are of the view that quality costs fall into the iceberg principle, in that only the

small portion that is above the surface, such as re-building costs, is visible. The part that cannot be seen because it lies below the surface however is likely to form a much larger proportion of the cost, possibly even as much as 90% of the total.

Universal standards of quality such as ISO 9000

ISO 9000 is a set of five worldwide standards that establish requirements for the management of quality. ISO 9000 is not a product standard but a standard for quality management systems. It is used extensively in the European Union. The standard is generic in that it applies to all functions and all industries. By 1992 more than 20,000 companies in Britain had adopted the standard and have become certified and over 20,000 companies from outside the EC have also registered. The Japanese have also mounted a campaign to get their companies registered. Documentation for ISO 9000 is onerous and requires the writing of quality manuals, the documentation of all relevant procedures and the writing of all relevant work instructions.

Benefits of certification

- Greater customer loyalty
- Improvements in market share
- Higher stock prices
- Reduced service calls
- Higher prices
- Greater productivity
- Cost reduction

There are also disadvantages of certification. As this chapter has consistently stressed, TQM is a philosophy and it is viewed that the rigidity of the ISO 9000 certification is contrary to this ideal. It tends to oppose the very objectives for establishing TQM in that the quality focus moves away from the customer towards the gaining of certification which becomes the primary objective. Likewise the responsibility for quality moves away from employees and back to the quality department responsible for obtaining certification.

Research in the construction industry has shown that there is no conclusive evidence one way or another to prove the benefit of certification in the construction industry[11]. However it must be borne in mind that quality systems such as ISO 9000 and TQM are not the same thing. In addition the criticisms that have been made of certification may relate not to the systems themselves but to the way they are being implemented.

Change management

One thing that will be obvious from reading this chapter is that the move to become a TQM company will involve change at all levels of the organisation. Change however will not happen effectively by itself; it needs to be managed. Managing change means taking control of the process and shaping the direction that the change will take. The management of change is a four-stage process[5]:

- Establishing the need for a change
- Gaining and sustaining commitment
- Implementation
- Review.

One of the important aspects in the effective management of change is that it must involve all members of the organisation. This is opposed to the current view of where change tends to be decided at senior management levels and then passed down to lower levels of the organisation with little explanation as to its rationale. Alternatively there is the mushroom approach whereby management simply throws some ideas to a situation they do not fully understand and wait and see what happens.

Even when change is managed there are barriers to it. These include cost, lack of time, employee perceptions, industry culture and lack of ability. Recognition and understanding of these barriers to change are an important step to overcoming them.

The methods of TQM

Any book on quality and TQM will include a vast array of quality and TQM methods. Most of these methods have been tested extensively in manufacturing industries but whether or not they are applicable to construction is not known. For this reason they are mentioned below but are not included in detail. Any reader wishing to study them further is referred to the references at the end of this chapter.

- Taguchi methods
- Failure mode and effects analysis
- Statistical process control
- Just-in-time.

How to implement TQM

In order to implement and maintain a system of TQM there must first be planning at the strategic level. Such a plan should include[10]:

Management of quality

- Quality definitions
- Total quality policy
- Total quality strategy
- Total quality culture.

Management of people

- All employees should understand their position and that of others in the organisation.
- Commitment
- Team work
- Education
- Open management.

Management of process

- Process design
- Process control
- Process improvement.

Management of resources

- Wealth generation
- Cost of quality
- Resource conservation
- Resource planning.

Kaizen[12]

One item which needs to be mentioned in relation to TQM is the Japanese concept of *kaizen*. *Kaizen* strategy is probably the single most important concept in Japanese management. The message of the *kaizen* strategy is that not a day should go by without some kind of improvement being made somewhere in the company. In Japan management is perceived as having two functions: maintenance and improvement. *Kaizen* signifies small improvements made in the status quo as a result of ongoing efforts. It is different from TQM in that it operates within existing cultures and rarely requires cultural shifts. Whether or not *kaizen*-type systems could be made to work in the West is not known.

Current research into TQM in the construction industry

One section of research in the field of TQM looked at certification. Research on the impact of the introduction of the quality assurance standards BS5750/ISO 9000 in construction-related organisations highlighted that there are very mixed views on the benefits of certification.[11] Quality certification was also examined in relation to architectural practices[13], and it was found that in relation to building contractors, architectural practices were slow to adopt quality assurance (QA) systems. The percentage of architectural practices with certified QA systems is in fact a very small percentage of the total.

Further work in this field[14] has established that certification may not be the best route to take and that possibly the most effective means of establishing a total quality culture and improving quality is for the company to develop its own quality improvement team. Work on quality control among contractors[15] in South Africa showed that contractors recognised the importance of quality systems particularly with regard to their use as a criterion for contractor prequalification. However in keeping with the work undertaken in the UK it was shown that many South African contractors found the formality of quality systems a problem and often failed to realise that it was possible to have quality systems that were based on the needs of their own organisations. Many of the contractors failed to realise that a quality system, to be effective, did not necessarily have to be formalised.

Another area which has been examined is the relationship between procurement and quality management[16]. This work is new and although it indicated that contractors may feel that design and build is the most appropriate format for addressing quality issues there is no conclusive evidence to support this.

Conclusion

In a recent lecture on process improvement one of the authors' students voiced the view that if something is not broke why try and fix it? In some ways this idea is the opposite of the TQM culture where it is believed that no system, however good it may appear, cannot be improved further. As this chapter has constantly stressed, TQM, possibly more than any of the other techniques in this book, requires a cultural shift for its effective implementation. Without this realisation it is highly unlikely that TQM will be achieved. For this reason this chapter has not tried to give details of how TQM systems operate but a more general view of the culture and environment required for successful TQM.

TQM is basically an attitude whereby customer focus is the main driver of the organisation. This is achieved within an organisation which is integrated and that is committed to continuous improvement. These are the three basics of TQM. Within these a multitude of systems are available and within those systems a multitude of techniques. Some systems of TQM are highly formalised. However there is no evidence to suggest that formalised systems are any better than bespoke systems developed in-house.

There can be no doubt that the construction industry is an enormously complex one and that as an industry it does have barriers that appear to prevent the development of a quality culture. Competitive tendering is an obvious example in that contractors who are forced into low prices as a means of securing work will always have to cut corners to stay within budget. Under such circumstances quality is an obvious casualty. Subcontracting appears another barrier. Whatever quality systems a contractor may have in place, these can become extremely difficult to implement when the entry of subcontractors into the industry is so easy and when the means of appointing them takes no account of the subcontractor's own quality. In addition, there is the difference in the level of management sophistication between the main contractor and the subcontractor[17]. The skill shortage makes matters worse.

These however tend to be circular arguments of the vicious circle variety. The only way that the circle can be broken is through a cultural shift, not only in the contractors but in the clients of the industry, the consultants and the subcontractors. It is the authors' view that this cultural shift has started and that once it gathers momentum the construction industry will change more quickly than has been anticipated.

References

1. Kelly, R. (1996) Heartache of a faulty house. *The Times* Wednesday 6 November, p. 42.
2. Rampsey, J. & Roberts, H. (1992) Perspective on total quality. *Proceedings of Total Quality Forum IV*, November, Cincinnati, Ohio.

3. Vincent, K.O. & Joel, E.R. (1995) *Principles of Total Quality*, Kogan Page, London.
4. Evans, J.R. & William, L. (1993) *The Management and Control of Quality*. West Publishing Company, Minneapolis.
5. Asher, J.M. (1992) *Implementing TQM – Small and Medium-sized Companies*. TQM Practitioner series. Technical Communications (Publishing) Ltd, Letchworth, Hertfordshire.
6. Deming, W.E. (1986) *Out of the Crisis*. Massachusetts Institute of Technology, Cambridge, Mass.
7. Bounds, G., Yorks, L., Adams, M. & Ranney, G. (1994) *Total Quality Management: Towards the Emerging Paradigm*, McGraw-Hill, New York.
8. Baden, H.R. (1993) *Total Quality in Construction Projects. Achieving Profitability with Customer Satisfaction*, Thomas Telford, London.
9. University of Cambridge. (1993–94) The financial impact of Japanese production methods in UK companies. *Research Papers in Management Studies*, **24**.
10. Choppin, J. (1991) *Quality Through People. A Blueprint for Proactive Total Quality Management*. IFS Publications, UK.
11. Hodgkinson, R., Jaggar, D.M. & Riley, M. (1996) Organising for quality. *Construction Modernisation and Education. CIB Beijing International Conference*, Beijing. On CD-ROM.
12. Imai, M. (1986) *Kaizen: the Key to Japan's Competitive Success*. McGraw-Hill, New York.
13. Emmitt, S. (1996) Quality assurance – more than a marketing badge. *Construction Modernisation and Education. CIB Beijing International Conference*, Beijing. On CD-ROM.
14. McCaffer, R. & Harvey, P. (1996) Total quality construction: a never ending journey. *Construction Modernisation and Education. CIB Beijing International Conference*, Beijing. On CD-ROM.
15. Rwelamila, P. & Smallwood, J. (1996) The need for implementation of quality management systems in South African construction. *Construction Modernisation and Education. CIB Beijing International Conference*, Beijing. On CD-ROM.
16. Rwelamila, P.D. (1995) Quality management in the SADC construction industries. *International Journal of Quality and Reliability Management*, **12**(8), 23–31.

17. Cheetham, D.W. (1996) Are quality management systems possible? *The Organisation and Management of Construction, Vol. 1: Managing the Construction Enterprise.* CIB W65 International Symposium, Glasgow, pp. 364–5.

Chapter 7
Supply chain management

Introduction

Construction management has often borrowed concepts from other management disciplines and the supply chain management concept has been adopted from mainstream management disciplines.

The application of the supply chain concept to construction has emerged in recent years as a management strategy to improve the performance of the industry. It has found favour particularly with the research community and various government policy making units worldwide. The application of the concept has had mixed usage among construction industry practitioners, although this appears to be changing, with ad hoc examples of industry participants such as clients, consultants, contractors, specialist subcontractors and major suppliers engaging in various forms of supply chain management behaviour, particularly in developing more long-term strategic relationships. Some notable applications by construction firms have been reported in the literature. There is very little known about the application of supply chain management in a widespread manner across the industry.

Supply chain management became a popular management tool and found favour in many industries relying upon manufacturing production and distribution functions and has been used with a high degree of success within companies. The fragmented nature of the construction industry and the perceived poor performance in productivity prompted many to look to significantly better performing industries such as those of the automotive, retailing and information technology sectors to adopt better management practices. One of the key drivers associated with this is the penchant to model the design and construction process as a manufacturing process.

There have commonly been two major directions in the consideration of the supply chain management concept in construction: a strategic procurement methodology and the logistics approach. Supply chain management was closely allied to the logistics concept in the early appropriations by construction researchers.

Construction product manufacturing companies have implemented supply chain management far more readily than other company types, in both tactical logistics and strategic procurement levels. The practical application of the concept within other industry types has been prolific and this appears to be driven by the widely held belief that supply chain management is easier to implement in process oriented industries rather than project based industries, given the tools and techniques that have been developed. The structural and behavioural characteristics of the construction industry appear to be a barrier to widescale implementation. Nevertheless the concept has potentially far reaching boundaries and the next decade may see changes to the way the supply chain is managed.

Definitions

There is much debate in the research community on the confusion regarding interpretations of the supply chain concept[1,2]. Much of this confusion arises because supply chain management is viewed from a number of perspectives. Over the past decade various definitions for supply chains and supply chain management have been proposed in the management literature[3–7], of which some general themes can be identified.

Some of the key themes identified in these definitions are:

- Improved customer value and reduction of costs
- Strategic management of the chain of relationships
- Synchronisation of information, product and funds flow
- Competitiveness, market forces and innovation

Christopher[3], one of the founders of the supply chain management concept from a logistics perspective, proposed that 'the supply chain is the network of organisations that are involved, through upstream and downstream linkages, in the different processes and activities that produce value in the form of products and services in the hands of the ultimate customer'. He went further to claim that 'supply chain management is the management of upstream and downstream relationships with suppliers and customers to deliver superior customer value at less cost to the supply chain as a whole'[3].

Poirier and Reiter[4], in their definition, offered the term 'system' to describe supply chains and thus proposed a concept of interacting components with interdependencies: 'A supply chain is a system through which organisations deliver their products and services'. Poirier and Reiter's model of supply chains was then developed on the basis of the chain as a network of interlinked organisations, or constituencies, that have as a common purpose the best possible means of affecting that deliver[4].

One of the key issues to this model that has been difficult to reconcile with the construction industry is that many construction organisations along the supply chain typically do not have a common purpose of procurement delivery. It has been acknowledged that in the construction industry each actor along the supply chain typically places a high priority on short-term project gains. It is probably for this fundamental reason that since construction clients have the most to gain through managing the supply chain for long-term benefits then they are the actors who ultimately need to take control[8].

There are a number of definitions that focus upon the physical distribution of products/materials along the supply chain. For example, Copacino[5], from a manufacturing industry logistics perspective, suggested that 'Logistics and supply chain management refer to the art of managing the flow of materials and products from source to user. The logistics system includes the total flow of materials, from the acquisition of raw materials to delivery of finished products to the ultimate users'.

There has been confusion and debate regarding the distinction between logistics and supply chain management for a number of years. It is now widely accepted that logistics and supply chain management are indeed two separate areas where logistics is subsumed by the broader field of supply chain management. As early as 1992, Christopher[3] made the distinction between logistics and supply chain management. In 1994 the Supply Chain Forum[9] developed the following definition: 'Supply chain management is the integration of key businesses processes from end user through original product suppliers that provides products, service and information that add value for customers and other stakeholders.' In 1998 the distinction between logistics and supply chain management was clarified by the Council of Logistics Management, a peak international body of industry and academic representatives[9]. This distinction between logistics and supply chain management was made in the construction literature in 1998[8].

Whether taking a logistics or managerial approach, underlying many definitions is a *strategic* perspective of the organisations that compose the entire supply chain. This strategic approach was explicitly highlighted by Ross[7]. He acknowledged that there were two levels with which to conceptualise supply chain management, namely the strategic and tactical; his text concentrated on the emerging strategic capabilities of the SCM concept: 'Supply chain management is a continuously evolving management philosophy that seeks to unify the collective productive competencies and resources of the business functions found both within the enterprise and outside in the firm's allied business partners located along intersecting supply channels into a highly competitive, customer-enriching supply system focused on developing innovative solutions and synchronising the flow of marketplace products, services, and information to create unique, individualised sources of customer value[7].'

Although the definitions identified in this chapter are developed for varying models and with different approaches, there are three underlying principles. Supply chain management is concerned with:

- Strategic perspective
- Customer value
- The effective economic organisation of firms along supply chains

With this in mind, London and Kenley[10] proposed the following working definition specifically for supply chain procurement for the construction industry: 'Supply chain procurement is the strategic identification, creation and management of critical project supply chains and the key resources, within the contextual fabric of the construction supply and demand system, to achieve value for clients.'

Figure 7.1 was an attempt to develop a model of the supply chain more representative of the major groupings of construction industry actors. Another important feature in the construction industry is that there are generic construction supply chains such as Figure 7.1 indicates; however each project can have unique procurement relationships that alter the chain dramatically. This is particularly relevant with contractors, clients and consultants and design and construct (design build), project alliance and build own operate and transfer contracts. Alternative procurement methods, that have included actors who have traditionally been located further down the chain, for example specialist subcontractors or manufacturers, can begin to play a particularly critical role in supply chain management. However these are emerging situations and attention on the success of such contracts as this belies the overwhelming fact that the vast majority of procurement modes with the vast majority of actors further down the chain have remained unchanged.

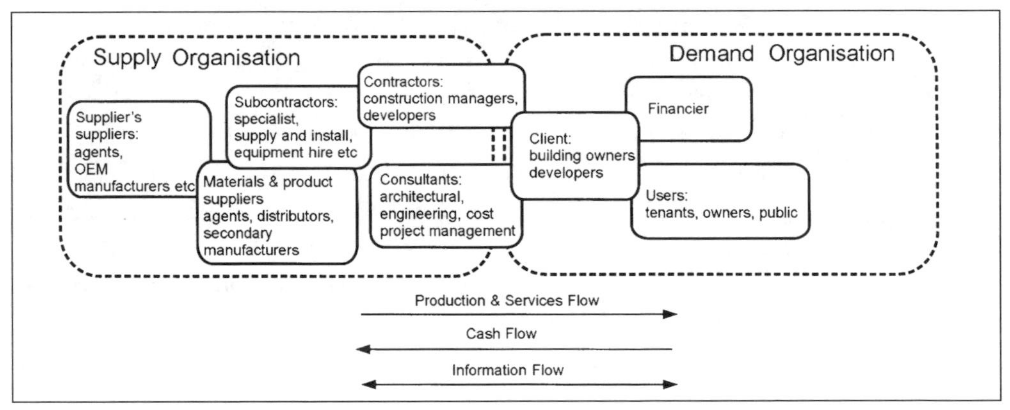

Fig. 7.1 Construction industry supply chains (adapted from Copacino[5]).

Goals of supply chain management

Governmental policy makers, such as the Australian Commonwealth Department of Industry, Science and Technology (DIST), have been attracted to the supply chain concept because of the problems of fragmentation and low productivity identified in the construction industry. Appropriating the supply chain motif is further rationalised by the successes of supply chain management in other industries.

Anticipated benefits listed by the Australian Commonwealth include:

- Faster response times
- Less waste
- Reduced inventory holdings
- Increased return on investment
- More effective information flow
- Lower cost
- More profit[11]

Clearly, construction innovation could be added to this list.

Charting the supply chain movement

The supply chain concept is part of an eclectic and developing hybridised field. It became an explicit area of research in the mid-1980s and originated largely from the two separate management streams of distribution and production, which merged into the field of logistics[12].

Since it became an identifiable area of research, the supply chain concept has been widened through the influence of other research frameworks. London and Kenley[13] identified the following four themes:

- Distribution
- Production
- Strategic procurement management
- Industrial organisation economics

Construction research involving the supply chain concept is also a relatively new field, having explicitly emerged in the mid-1990s. Similar to the mainstream management literature, it is evolving with corresponding perspectives from the theory of production, distribution and strategic procurement. Significantly, there has not been any construction industry research merging the supply chain and industrial organisation fields, as found in other industry studies by Nishiguchi[14], Ellram[6], Hines[15], Harland[16] and Lambert *et al.*[9].

Figure 7.2 charts some of the more significant supply chain events, models and definitions against these four themes for the two decades between 1980 and 1999. The italicised text indicates the research related to the construction industry.

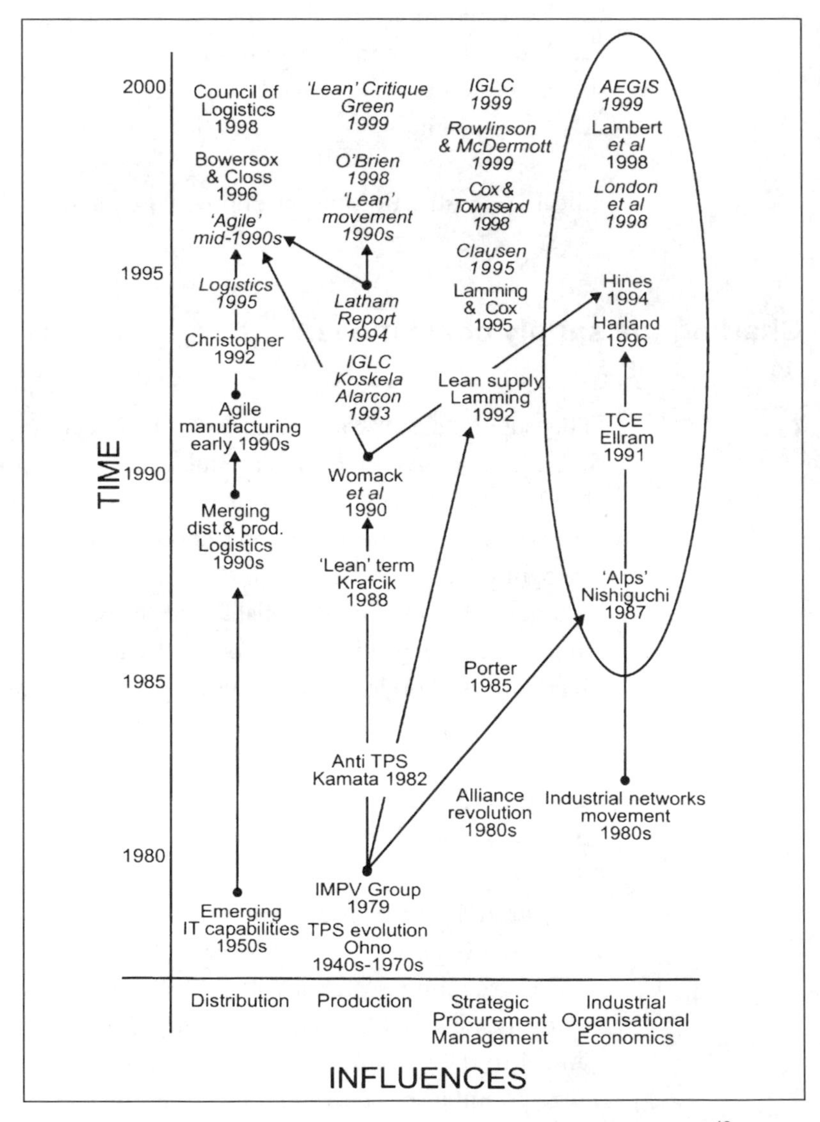

Fig. 7.2 Chart of supply chains streams of research 1980–1999[13]

It is worthwhile to understand some of the research and associated examples from other industries, as many of the concepts are applicable. However this should be done with caution, as supply chain management is first and foremost a concept that can only be applied when business relationships between actors along the chain are fully understood. Business relationships between customers and suppliers largely rely upon the relative degree of power that suppliers have within their own market. Once power relationships between customers and suppliers are understood then a variety of supply chain management tools and techniques can be applied.

Distribution

Distribution: mainstream management

Supply chain management has often been associated with the management of the physical distribution of products from raw material through manufacturing processes to 'point of sale' for the end product.

Christopher[3] has been considered one of the pioneers of the logistics and supply chain movement. Borrowing from Porter's value chain concept[17] he moved the perspective of materials management from a tactical low level task in the organisation to a strategic management concept that supports customer focus and creates competitive advantage.

Such research has emphasised the development of integrated supply chain processes to support planning and coordination of complex supply chain systems for efficient and timely movement and storage of products and/or materials[18]. For example, modelling of these systems has involved mapping of time and cost resources and considering such concepts as *time-compression* and *just-in-time* relative location of stock and warehouse management, transportation analysis and optimisation models to improve logistics performance.

This approach has often relied upon hard systems methodologies to model, forecast and predict the product, information and funds flow. Research methodologies have been characterised by both case studies and simulation models for *fast moving consumables*. In the retail industry technological innovations such as electronic data interchange and vendor managed inventory have harnessed the capabilities of information technology to radically alter the flow of information and to be more responsive to changing customer trends. This flexibility in the chain, in response to customer demand, has been the cornerstone of the *agile manufacturing* concept.

Distribution in construction: materials flow and just-in-time

Construction researchers have also applied the supply chain management philosophy to materials flow, seeking to establish the relationship between site productivity and improved materials management systems[19–22].

There is little doubt that construction materials manufacturers have applied logistics concepts to their manufacturing processes. Large materials suppliers, for example glass suppliers, have applied the just-in-time delivery philosophy.

Agile construction has been taken up by some construction researchers, who argued that *lean practices* and benchmarking would be an essential ingredient in achieving the target of a real cost reduction of 30%[23].

The concepts of agile construction and lean construction are blurred, with some claiming that there is a difference[24] and others using the concepts interchangeably[23]. Lean construction seems to have more support within the construction research community and is discussed in the next section on production.

Merging production and distribution

Prior to the 1960s the fields of production and distribution were fragmented, eventually evolving into two identifiable streams in the early 1980s. In the last decade the research area of integrated logistics management has subsumed both fields[12].

There has been much debate attempting to distinguish the fields of logistics and supply chain management. This distinction was clarified by the Council of Logistics Management, a peak international body of industry and academic representatives[9]. It was accepted that supply chain management was more than simply logistics and operational issues and that strategic supply chain management subsumed logistics. The distinction between logistics and supply chain management was made in the construction literature in 1998[25]. The majority of construction production researchers have tended to focus upon the production and logistics debate when discussing supply chains.

Production

Lean production: mainstream management

Production theory, particularly *lean* production, has been another major influence on the supply chain movement. One of the seminal texts is *The Machine that Changed the World*[26], which resulted from an international benchmarking study, associated with the International Motor Vehicle Program at MIT. Womack *et al.*'s study described and analysed the method of production termed lean production, best exemplified by the Toyota Production System, and pioneered by the Japanese executive, Ohno. The term *lean* was actually first coined to describe this system by Krafcik[27].

As discussed by London and Kenley[13] the epistemology of lean production is posed against *craft* and *mass* production. Craft production was based upon the notion that manufacturers of complex products required skilled labour within a collaborative environment, supported by a system of apprentice–journeyman–master and Craft Guilds, etc. This gave way to mass production whereby unskilled labour could perform tasks designated and instructed by managers. Lean production is often considered a reaction by the Japanese to mass production.

One of the most important lessons from the lean approach is that it attempts customisation of high volume production, to provide customers with exactly what they want at the time they want it. To achieve this end the lean approach is characterised by improving flexibility, reducing waste and improving flow along the supply chain. The flow is improved through management and control of each actor along the supply chain. This largely relies upon some form of 'total' integration from raw material supplier to various subcontractors who supply materials or components to the manufacturer.

Lean construction

The lean construction movement has, from 1993, led much discussion on supply chains through the International Group for Lean Construction annual conferences. Lean construction evolved from lean production, a developing field that is centred primarily upon a production philosophy for construction. In so doing, key protagonists have explored workflow and conversion processes, waste reduction and efficient use of resources, through lean project management, lean supply, lean design, lean partnering and cooperative supply chain management[28]. The central themes have been eliminating waste and improving workflow in the construction industry.

To date, much of the construction literature has applied the lean concept without contextualisation, for example, without the detailed empirical exploration of market structures that underpin the construction environment. Those researchers in the lean production field have understood this important issue. The contextualisation of lean production that supports lean thinking has been provided through organisational and industrial organisational economic descriptions of the automotive and electrical industry supply chains[15, 29–31]. This understanding of the organisation of the supply chain is an important part of the philosophy of lean thinking.

An example of the manner in which the nature of the underlying economic organisational market forces impacts upon supply chain management in both materials and workflow was highlighted in an Australian case study of small to medium sized construction firms exporting structural steel prefabricated housing[32]. This example of a supply network of some 20 firms was identified in the supply organisation with a variety of procurement relationships, ranging from strategic and critical to non-strategic and non-critical.

Ethical considerations

In recent times the value to the construction industry of the lean thinking dogma and rhetoric[33] has been questioned. Green challenged the narrowly defined instrumental rationalist approach currently undertaken in the movement. He aimed at introducing literature that provided evidence of the human cost of lean methods in Japanese industry, including repression of independent trade unionism, societal costs (pollution and congestion) and regressive models of human resource management[31, 34]. Kamata[34], provides a personal account of life as an assembly-line worker inside the walls of a Toyota plant, the physical exhaustion he experienced meeting impossible production targets, the army-like treatment of employees, the need for conformity and tight surveillance, etc. Green[33] argued that 'whilst the lean rhetoric of flexibility, quality and teamwork is persuasive, critical observers claim that it translates in practice to control, exploitation and surveillance'.

Lean proponents responded, defending their movement with the argument that it is based upon a long history of production management thinking, particularly the physics of production[35] and that lean thinking simply offers a new way to organise production.

Tiered supplier networks

Lean production implementation by large producers in Japan would not have been possible had it not been supported by highly organised governance structures in the supply chain. Supply chains were organised into hierarchical clusters of tightly tiered structures of subcontracting firms; these clusters were known as *keiretsu*[15, 30]. Lean construction researchers, in their quest for production efficiency, in many cases have forgotten that organising and controlling the market on a very wide and deep scale was instrumental in lean implementation.

Nishiguchi developed an historical description and analysis of the Japanese subcontracting system. He provided empirical evidence of the development and organisation of industry from the 1920s through to the present day, highlighting the underlying structural characteristics of markets and the evolution of the subcontracting interorganisational relationships.

Japanese economists have typically debated the nature of Japan's subcontracting small enterprises from two perspectives. The first position relies upon the dualist theory which holds that 'big businesses accumulate their capital by exploiting and controlling small businesses which have little choice but to offer workers low pay under inferior working conditions'[31]. The prosperity of Japanese industries, particularly in the automotive and electronics sectors, lies with the sacrifice of many subcontractors. The core dualist theory suggests that economic agents, either workers or firms, located in different segments of the economy are treated unequally, regardless of their objective worth.

The second position emphasises the 'vitality, dynamism and innovativeness of small businesses that have responded flexibly to the needs of clients and markets'[31]. Nishiguchi[30] along with Sugimoto[31] attempts to reject the dualist theory, claiming that Japanese business is more complex. Nishiguchi presented empirical evidence of sustainable growth and high asset specificity of the small to medium sized subcontractor firms within lean supply. He also showed that union membership in Japan has remained the same and that interscale wage differentials between large and small firms was not as marked as some suggest.

Sugimoto[31] concluded that both positions exist and that the variation in value orientations and life style of workers is dependent upon the extent of control of the small businesses by the larger companies at the top of the hierarchy. Those who tended to diversify their connections were less controlled and more innovative, participatory and openly entrepreneurial.

Construction production

Alternative models to that of lean construction have been developed by researchers who also used production theory. This has its origins in the construction research community reconceptualising the construction industry as a 'manufacturing process', which was well supported in the UK government industry reports such as the Latham and Egan reports.

One approach, the generic design and construction process protocol, treats the industry as a production process[36]. This research described the industry as a single process map for all phases by adopting the manufacturing model of new product development.

Workflow

An alternative view of construction production theory which focuses more directly on the supply chain concept was also offered by O'Brien[37], who was concerned with materials flow, and also raised important questions for workflow. He investigated the production and inventory decisions of multiple firms within the construction supply chain. He indicated that any managerial philosophy, such as just-in-time, applied to one site for one project in the construction environment, is problematic due to the temporary nature of project organisations.

O'Brien[37] makes a further contribution with a systems view of the construction production supply chain, identifying that supply chain management offers the potential to optimise supply chain cost performance. By borrowing and modifying production manufacturing capacity cost models, he investigated[17] firms to identify how capacity constraints of subcontractors and suppliers affect the costs associated with construction project schedule and scope changes. The work forms the foundation to develop models for supply chain performance.

Strategic procurement

Mainstream management

A strategic perspective of the supply chain concept emerged in the 1980s which subsequently evolved into *strategic procurement*[3, 7, 17]. Typically this involved positioning a firm competitively in the marketplace by developing appropriate sourcing and management strategies for suppliers. Porter[17] developed the concept of the value chain as a tool for firms to improve competitive advantage in an industry. Further to this is the concept of strategic procurement management, which is the development of an external sourcing and supply strategy designed to maintain a sustainable position for that organisation in the total value chain.

Lamming and Cox[38] identified the importance of supplier development through allied business partners and strategic collaborative partnerships to enable lean production to take place. They termed this lean supply. In contrast to the accepted view of control in the lean supply chain, Lamming[38] suggested that achieving lean supply is a complex matter because of the nature of competition in markets as the suppliers are involved simultaneously in several other chains. Jealous guarding of expertise cannot be maintained in the lean enterprise, as it requires trust between firms.

Strategic procurement is much wider than the lean movement. It is a concept applicable to all firms and not simply those involved in production and manufacturing. A significant part of strategic procurement is concerned with business alliances. Co-operation among firms has grown rapidly since the early 1980s as alliances have proliferated in one industry after another[39].

Ross[7] acknowledged two levels to conceptualise supply chain management, namely the strategic and tactical, and his research concentrated on the emerging strategic capabilities of the supply chain management concept:

> 'Supply chain management is a continuously evolving management philosophy that seeks to unify the collective productive competencies resources of the business functions found both within the enterprise and outside in the firm's allied business partners located along intersecting supply channels into a highly competitive, customer-enriching supply system focused on developing innovative solutions and synchronising the flow of marketplace products, services and information to create unique, individualised sources of customer value'.

Strategic procurement in construction

Strategic management was a relatively new field in construction in 1991, with little literature available[40]. The intervening decade has not seen a growth in this area, but rather a growth in the research related to the management and procurement for the individual project. The strategic management of interorganisational relationships is still relatively new, with a growing focus on project alliances. Early research into such concepts as strategic alliances, serial contracting, multiple project delivery, organisational design, vertical integration and supply chain procurement[41] are indications of a growing awareness of strategic organisational management in construction supply chain research[42].

Supply chain integration and control

Cox and Townsend[41] proposed the critical asset and relational competence approach to construction supply chain management model that relied upon clients controlling the supply chain. The authors advocated for clients to understand the underlying structural market characteristics of their own construction supply chains and to develop contingent approaches to procurement based upon this understanding. They considered the UK construction research based upon lean and agile manufacturing inappropriate, because it lacked contextual understanding of the construction industry. They even suggested that:

> 'It is our view that if the Latham report, and the somewhat naive research industry into automotive partnerships and lean and agile manufacturing processes that it has spawned, had devoted more time to analysing and understanding the properties of the unique supply chains which make up the complex reality of the UK construction industry a greater service might have been done to value improvement in construction.'[41]

Strategic alliances

Other researchers have conducted similar case study research on strategic procurement and supply chain management. Olsson[43], through a qualitative case study on supply chain management of a Swedish housing project driven by Skanska and IKEA, highlighted that a conventional construction approach was found to be too expensive to meet particular client demands.

Consistently, researchers have concentrated upon a small group of firms and the supply chain management concept.

Vertical integration vs SCM

Clausen's Danish study[44] also focused upon the key firms in the main construction contract as they evaluated a government programme, where the 'government', acting as a large client, intended on improving productivity and international competitiveness in the construction industry. The central argument to the programme was the need for vertical integration of the different actors and their functions in the construction process[44], with the premise that key actors in the process should be involved in strategic decisions from the outset.

Clausen[44] determined that the programme was much less successful than anticipated because there was discontinuity in the supply of projects to the consortia; firms were concerned about the financial risk of committing their resources to a single client. The conclusions suggest that the degree of uncertainty in supply of projects and the inherent risk for firms involved is a very important factor in supply chain management.

The interplay between supply and demand, the balance of power or control and incentive has been considered by others in the form of serial contracting and multiple project delivery[42]. Although authors suggest the importance of understanding the entire scope of the supply chain, the supply chain is often still perceived as the contractor's supply chain[45]. London and colleagues[8, 10] suggested that the client is the more likely proponent and beneficiary for the management of the supply chain.

In all these models, construction supply chains are viewed from the perspective of the single organisation and its ability to control other firms. There is still however a dearth of empirical research to address the supply chain across the breadth of the industry, thus advancing the debate toward understanding, describing and analysing the structure and behaviour of supply chains.

Many authors in strategic procurement have moved the debate regarding supply chains with respect to the need for the development of appropriate relationships, the problems of unreliable supply, the different degrees of control between firms and the difficulties due to the temporary nature of a project based industry. However these are all characteristics of the real world construction industry. The approach discussed in the next section is related more to understanding the structural and behavioural characteristics of markets and the relationship between markets, firm behaviour and the supply chain concept.

Industrial organisation economics and supply chain concepts

The industrial organisation methodology deals with the performance of business enterprises and the effects of market structures on market conduct (pricing policy, restrictive practices and innovation) and how firms are organised, owned and managed[46]. The most important elements of market structure in these models refer to the following factors:

- The nature of the demand (buyer concentration, number and size of buyers)
- Existing distribution of power among rival firms (seller concentration, number and size of sellers)
- Government intervention
- Physical structuring of relationships (horizontal and vertical integration)

The role of the industrial organisation model is to give substance to the traditional neoclassical abstract concepts of market types.

Transaction cost economics and supply chains

Ellram[6] took an industrial organisational perspective to the supply chain concept, although from a single organisation's ability to manage the supply chain. She suggested types of competitive relationships that firms undertake ranging from transaction, short-term contract, long-term contract, joint venture, equity interest to acquisition. These involve increasing commitment on the part of the firms. She described the key conditions under which supply chain management relationships are attractive according to an industrial organisation perspective. The main thrust was that supply chain management is simply a different way of competing in the market that falls between transactional type relationships and acquisition and assumes a variety of economic organisational forms[6].

This was one of the first discussions to explore the implications of Williamson's transaction cost economic theory and industrial organisation economics related to supply chain management. Situations conducive to supply chain management included:

- Recurrent transactions requiring moderately specialised assets
- Recurrent transactions requiring highly specialised assets
- Operating under moderately high to high uncertainty

Such prescriptions should be considered with caution; arguments designed to prove the inevitability of this or that particular form of organisation are hard to reconcile, not only with the differences between the capital and socialist worlds, but also with the differences that exist within each of these.

However, the transaction cost economics theory has just as many critics as supporters. One of the main criticisms is that it has tended to assume a market and hierarchy dichotomy. Theorists have found it difficult to explain contractual relationships between firms where clearly the transaction costs were high and yet firms did not vertically integrate. There are a variety of institutional arrangements between the two extremes of market versus hierarchy, which do not fall neatly into the transaction cost model and clearly demonstrate that markets are not the only way prices are co-ordinated. However, there is potential for future research relating transaction cost economics to the supply chain movement for the construction industry. Transaction cost economics tend to focus upon individual firm-to-firm contractual relationships whereas supply chain theory aims to understand many interdependent relationships as the unit of analysis.

Describing supply chains

One of the significant contributions by those using concepts derived from the industrial organisation economics literature is the attempt to describe and analyse the structuring and interdependency of relationships in the system of supply chains. New[1] noted that the development of the idea of the supply chain owes much to the emergence from the 1950s of systems theory and the associated notion of holistic systems. There are many variations to systems theory but at the core is the observation that a complex system cannot be understood completely by the segregated analysis of its constituent parts.

Selected supply chain research published in mainstream management literature has studied the complex system of supply chains through interorganisational structure. These are important models that merge the field of industrial organisation and supply chain theory[9, 15, 16, 30, 47]. There are two main epistemological differences between this stream of research and the others discussed thus far. Firstly, industrial organisation supply chain research tends to be primarily descriptive in the first instance, rather than prescriptive, and is more about supply chains rather than supply chain management. New[1] explains the twin dichotomy between research on 'supply chains' versus 'supply chain management'. Secondly the unit of analysis is not the individual firm, nor the individual relationship, but rather an aggregation of firms and relationships.

Supply chains for lean systems

Although Hines[15] and Nishiguchi[30] are clearly advocates of the lean system of supply, some of the more significant contributions of their research were the descriptions of the historical, organisational and economical structure of the Japanese system of supply across automotive and electronics industries. In many ways this has provided a richer picture of lean production and supply chains than other writings of the apocalyptic posturing of the field's success.

Typically, suppliers are categorised and organised into either specialised subcontractors or standardised suppliers, based upon the degree of complexity of the supply item. It is within the specialised subcontractors that the pyramidal Japanese subcontracting system or the concept of clustered control lies. This system has traditionally been described as a pyramid with an individual assembler corporation at the top and successive tiers of highly specialised subcontractors down the chain, increasing in number and decreasing in organisational size at each progressive stage. This represents a single company network encompassing all the relevant tiers necessary to produce the end product. This suggests a closed system; however, in reality, first tier suppliers supply to many assemblers across the industry[30]. This led to the Alps structure of supply chains, a series of overlapping pyramids resembling mountain alps across an industry. Hines[15] enlarged the industry specific view to look at the wider economy and suggested that rather than this closed rigid system, the Japanese subcontracting system is moving more towards a structure of interlocking supplier networks. In this system, many firms supply more than one industry sector and potentially operate in different tiers, for example the electronics suppliers operate in a number of sectors.

Typically associated with each tier are supplier associations, which are a mutually benefiting group of a company's most important subcontractors, brought together on a regular basis for the purpose of co-ordination and co-operation as well as to assist all the members. Major materials-supplier corporations are typically sourced directly by the large assembler corporation and through strategic procurement; sourcing alliances are dealt with separately from the pyramid system. The supply of the material is provided to appropriate subcontractor tiers for the manufacture of components[15].

Hines[15] developed a further technique for understanding a particular supply chain appearance at an overview or industry level and termed it physical structure mapping. Within this technique, the number of firms involved at various stages in the production process in the supply chain is related to tiers of suppliers. Subsequently, a secondary industry map is developed where the tiers are mapped against the value-adding process, or more strictly to the cost-adding process.

Mapping supply chains

In 1998 Lambert *et al.*[9] also provided insights for mapping supply chain structure. They claimed, quite simply, that there are three primary structural aspects of an organisation's supply chain structure:

- ❑ Members of the supply chain
- ❑ Structural dimensions
- ❑ Types of process links

With regard to structural dimensions there are three critical dimensions:

- Horizontal structure refers to the number of tiers across the supply chain, which is in effect the number of different functions that occur along the supply chain and indicates the degree of specialisation.
- Vertical structure refers to the number of suppliers and customers represented within each tier. This reflects the degree of competition amongst suppliers.
- Horizontal position is the relative position of the focal company within the end points of the supply chain.

Lambert *et al.*[9] developed for the supply chain structure of an organisation a generic map, a complex network of suppliers and customers arranged in successive tiers from the focal organisation. In many ways this model suggests methods of strategic procurement. However, the importance of this model for the industrial organisation debate is the inclusion of a number of empirical case studies indicating the structure of different supply chains and the interconnection between a number of focal organisation's supply chains and the resultant networks of supply.

Supply network

Harland[16] widened the industrial organisation of supply chains debate, by suggesting the term *the supply network* as a means for capturing the full complexity of the firms involved through a more holistic view of the process. A supply network can be defined as 'a number of entities, interconnected for the primary purpose of supply of goods and services required by end customers'. These entities may be engaged in long-term relationships, but the boundary of the network is ultimately ambiguous. In reality, a spectrum of supply relationships exists, ranging from tight long-term to loose short-term relationships. The term supply network seems to be gaining increased acceptance in the literature[48].

One of the pioneers of the supply chain management concept, Christopher[3], understood that, in reality, supply chains and the supplier markets are quite complex and network-like. He offered the following definition of the supply chain:

> 'The supply chain is the network of organisations that are involved, through upstream and downstream linkages, in the different processes and activities that produce value in the form of products and services in the hands of the ultimate customer.'

Much of the supply and industrial network literature builds upon the industrial networks movement of the 1980s[49]. This body of research has tended to suggest that close-knit interorganisational networks produce superior economic performance and quality, and that there should be a move away from the large, vertically integrated firms[50].

Construction industrial organisation

The role of the supply chain concept in construction will soon move beyond the rhetoric that it is a management tool to improve the performance of the industry. Future research may include optimisation of supply chains and will enable more credible discussions of advantages of different types of networks, clusters or chains.

Some studies have widened the perspective and have introduced industrial organisational concepts, for example vertical integration[22, 44], subcontractor/contractor dependence and the 'quasifirm'[51] and buyer concentration or pooled procurement[52]. There is no shortage of construction supply chain research that is action, applied or case study in orientation. Much of this empirical work is oriented to the project as the unit of analysis. There is a lack of work that approaches the research problem from a wider industrial context. A deeper and more detailed understanding of industrial organisation theory and supply chains would further this debate and has been initiated[8, 53].

Projects: multi-market models in supply chains

An explicit examination of a construction project through a supply chains lens soon reveals that there are multiple markets operating within a single project supply chain. There are numerous transactions, numerous supply chains and numerous market models on individual projects.

In 1998 London *et al.*[25] did such an exploratory study of the type of markets operating across the supply chain for domestic aluminium windows in Australia. Borrowing from Hillebrandt's[54] method of assessment of the type of markets for contracting, which was dependent upon the method of selection of contractor, a matrix was developed that mapped the markets of a number of firms in the supply chain. The basis of market characteristics and types was the microeconomic four-market model.

Microeconomists use the four-market models to describe the degrees of competition within the mixed market system including perfect competition, monopolistic competition, oligopoly and monopoly. The four models rely upon three key elements that have been observed of industries:

- The number and relative size of firms in the industry
- Whether firms supply identical or slightly different products
- The ease with which firms can either enter or leave the industry

The structure of the construction industry is constantly changing. Different market models operate at different times in the industry related to the individual characteristics of the construction project. Hillebrandt[54] identified that the different market models are sensitive to the various tender selection models used for individual projects in the industry. Her conclusion was that prior to a tender being awarded there are a number of different types of markets operating dependent upon how the client chooses to approach firms to provide a bid for the project. This literature focused upon the role of one key participant and the major transaction in the supply chain: the contractor and the construction.

Further research is currently being conducted by London *et al.*[25] to determine the extent of consistency in structure and behaviour in the project construction supply chains for multiple supply chains. At this stage it is unknown whether or not there are patterns arising across supply chains in the construction sector.

London *et al.*[25] suggested that supply chains for construction projects can be quite complex in nature and should be managed to account for this complexity. There are a number of tiers similar to any production process. However there are layers of markets at each tier and markets are often unique to a project. Figure 7.3 maps a

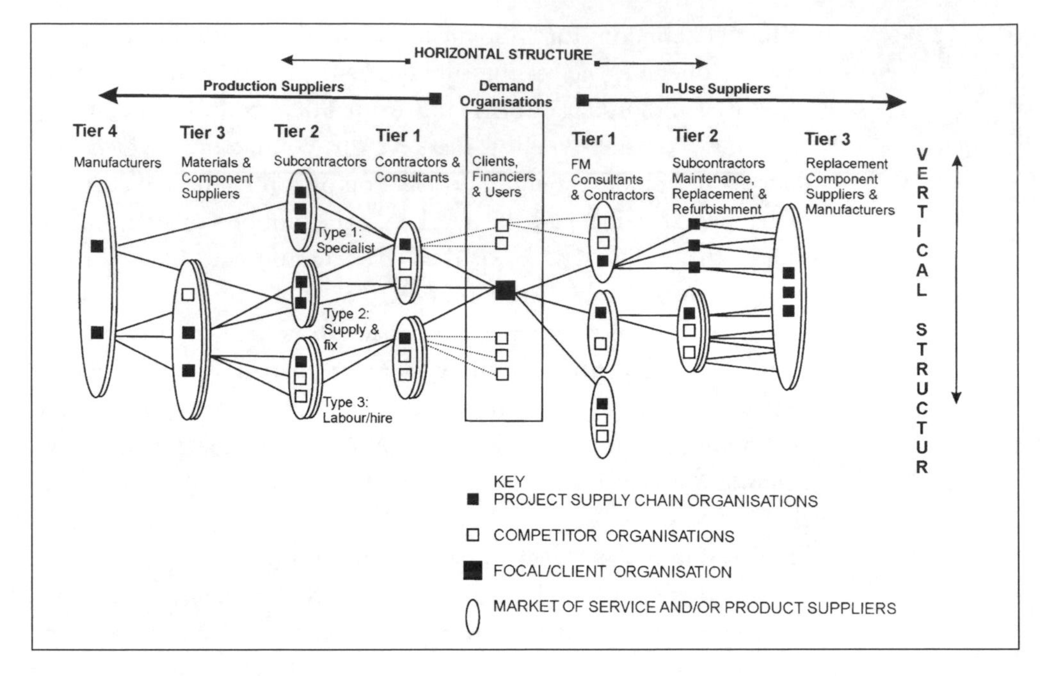

Fig. 7.3 Map of supply chains for building life cycle (adapted from Lambert *et al.*[9]).

generic supply chain structure for the building life cycle. It includes various suppliers involved in the management and maintenance phases. The importance of examining these characteristics is that supply chain management involves strategic decisions and strategic behaviour regarding procurement. Strategic decisions are influenced by forces that are operating both external and internal to firms. The type of market impacts upon the nature of competition amongst firms and therefore directly impacts upon the degree of power that upstream firms have when procuring from the suppliers in that market.

The AEGIS[55] model for the building and construction industry cluster also contributes to the development of the wider industrial organisation perspective. It discussed the industry as 'chain of production' and conceptualised the industry through five main sectors:

- On-site services
- Client services
- Building and construction

- ❑ Supplies and products
- ❑ Fasteners, tools, machinery and equipment

Existing statistics were used to describe the sectors in terms of industry income. However the authors note that this is contrived, as sufficiently detailed data are not available. The major firms are organised and listed according to groups, suppliers, project firms and project clients. The discussion focuses upon general information about size and turnover and addresses a market view of some key markets and the major players in the key markets, but does not seem to address the firm or project level of supply chains.

There is a need to develop this further and explore the explicit interfirm supply chain relationships on projects within the context of the firm and market. The firm and market level of analysis lies within the field of industrial organisation economic theory.

Government public policy, particularly competition policy, should be informed by observing the current state of the supply chain and distilling certain properties about industrial organisation. Until we are able to describe the vertical and horizontal relationships between firms and understand interdependencies at a firm level in relation to the market level, it is difficult to compare the long-term impact upon changes to the relational position between firms. In a global economy this may also have implications for competitiveness, sourcing, monitoring and traceability of products and materials. Specifically it will assist in understanding new players in the chain as e-commerce becomes more and more significant, for example dedicated supply procurement managers or transaction organising companies.

Networks, chains, clusters and constellations

A more strategic approach to procuring and managing the supply chain was identified in a study of very small Australian construction firms who have formed a network of alliances to penetrate international markets[32].

In construction, alliances have been suggested as a form of governance structure to solve procurement issues as there is a need for improved interorganisational relations between organisations[56]. In some research the discussion has suggested that alliances have potential to solve the construction industry problems, for example to reduce costs or to improve quality. The literature regarding construction alliances is quite sparse; however, attempts have been made to describe different types of alliances. Love and his coauthors[56] described the 'learning alliance' for construction and developed a conceptual framework for developing an ideal learning model for a successful alliance. Learning alliances are but one type of alliance. Gomes-Casseres[39] suggested there are three main motivations that characterise an alliance: learning alliances, supply alliances and positioning alliances.

Project alliancing has found favour in the construction industry in recent years. In the project alliance the contractual relationship between key firms on a project, formed specifically for a project and then when the project is completed the temporary organisation is disbanded, there is no contractual relationship binding the parties. The project alliance has a contractual relationship tied to an individual project and is largely short term in focus and includes the client in the relationship.

The case study by London[32] on the network of alliances differs from the project alliance perspective. The perspective of the network of alliances taken in this study is that the relationships are formed for more than one project and have a longer-term perspective. It could be suggested that it is difficult then to distinguish between this type of alliance and the construction joint venture for projects, where firms join together in a contractual relationship to develop a bid for a project. This type of relationship is aligned with this research and at times there may be a blurring of definitions. However, similar to the project alliance, the construction joint ventures can in many cases be short term in focus and typically narrow in long-term joint business strategy. The firms typically have little penetration into each other's business.

The case study indicated the existence of all three types of alliances. In some cases all three elements occur within the one alliance. The focus of this study is not specifically on the type of alliance or a single alliance, rather it is on describing the context of the constellation of alliances, that is the relationship between the network of alliances and the strategy for the group. There were alliances related to the construction procurement and instances of projects, however they are simply one small part of the entire constellation. The governance structure is much broader than the usual individual construction project approach.

The network of firms joined by various types of alliances evolved over some 20 years as the key decision makers learnt and reacted to the market environment and involved approximately 20 firms. However the firms were classed as small or even micro – employing in some cases two or three people. The constellation was structured according to the strategy to penetrate the international market for affordable housing through using an innovative building product system of prefabrication. As the case study revealed, the affordable housing market is closely allied to the search for markets for innovative building products using waste material. The case study demonstrates some of the conflicts, constraints and issues that concerned the actors in small companies that are involved in the process. Figure 7.4 shows the structure of the constellation in late 2000.

Tools and techniques

Supply stream classification

Most strategic sourcing approaches have emerged in practice through developing a customised strategy for each supply chain group or stream. An organisation may have many different supply chains, however some general classifications have enabled a shift in focus. The shift has been from that of simply procuring products and/or services to managing the acquisition and the use of the purchases to ensure that the organisation maximises the value of its investment. The supply stream classification principles apply to all procurement environments.

The following classification and development guidelines suggested by Riggs and Robbins[57] have been developed to assist senior managers to categorise purchases and to determine their supply strategy requirements.

Each supplier consolidation and procurement effort should begin by answering the following two questions:

(1) Does the product/service relate directly to the way an organisation creates value for its customers?

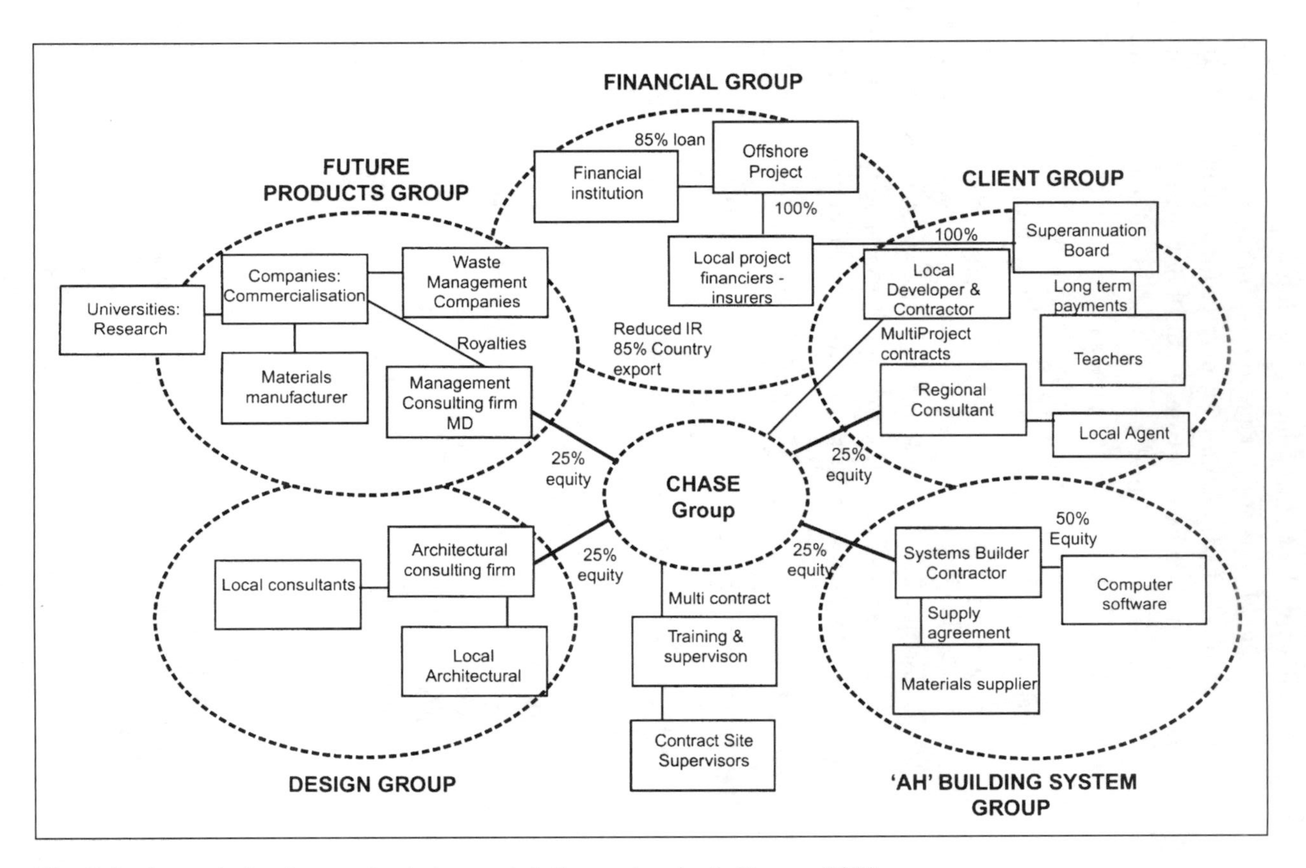

Fig. 7.4 Case study of a supply chain constellation: network of alliances 2000

(2) Does the market solution require a customised or standard product/service?

Riggs and Robbins[57] continue with their explanation by developing a quadrant matrix as follows:

- Quadrant 1: non-core/customised purchases
- Quadrant 2: core/customised purchases
- Quadrant 3: non-core/standard purchases
- Quadrant 4: core/standard purchases

Quadrant 1: non-core/customised purchases: typically supply managers sourcing these products or services search the market for the best, specialised, lowest cost, most efficient and most effective specialised solutions.

Quadrant 2: core/customised purchases: supply managers search the market for distinctive products, services and technologies through suppliers that are leaders in the distinctive market segment.

Quadrant 3: non-core/standard purchases: supply managers source and search the market for turnkey low-cost solutions wherein the supplier frequently manages the entire provision stream in addition to supplying the specific product or service; they may include office suppliers, maintenance tools, leased cars, etc.

Quadrant 4: core/standard purchases: supply managers search the market for distinctive market leaders with established positions and demonstrated competitive advantage in their products and services.

Cox and Towsend[41] gave a number of examples through their case studies on strategic procurement in construction.

London[53] reports on the procurement practices in the construction supply chain and gives an example of a matrix classification system that has been in use by two different major manufacturer and materials suppliers to the Australian construction industry for over a decade. The following matrix was developed by London to reflect three main criteria in supply positioning:

- Actual procurement patterns are based upon the behaviour of these firms

- It reflects the structural market characteristics of the firms supplying products and/or services
- It reflects the relative importance of the supplier's product to the customer

It differs somewhat from the previous matrix as it is based upon empirical observations of the industry, is not normative and focuses upon positioning based upon risk and expenditure to the buyer.

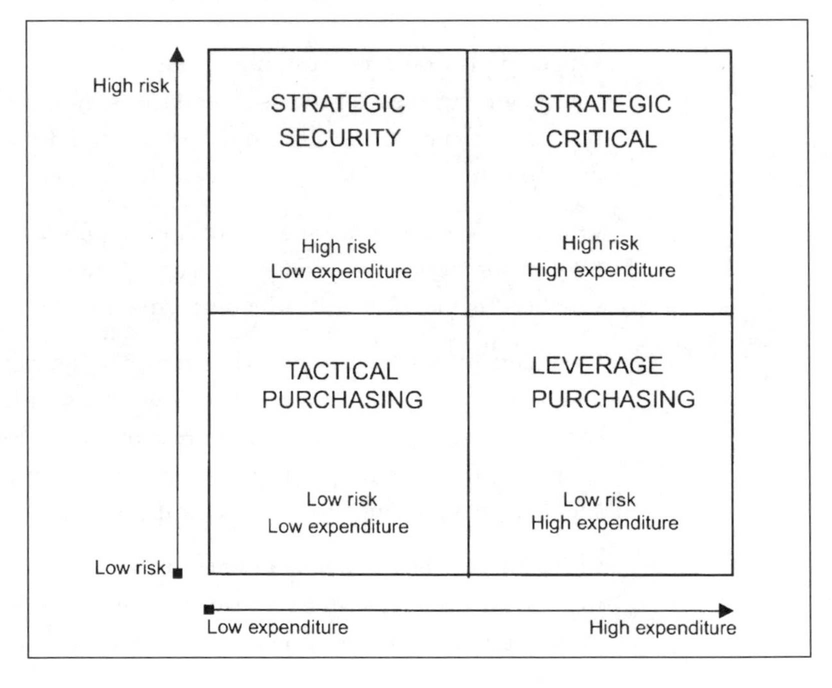

Fig. 7.5 Supplier procurement classification based upon risk and expenditure.

Supplier co-ordination and supplier development

Supplier co-ordination refers to the activities made by a customer to mould their suppliers into a common way of working, so that competitive advantage can be gained, particularly by removing intercompany waste. This type of co-ordination would involve areas such as working to common quality standards, using the same paperwork system, shared transport and employing intercompany communication methods such as EDI.

Supplier development refers to the activities made by a customer to help improve the strategies, so that suppliers can plan their processes more effectively, as well as the customer offering specific assistance to the suppliers in areas such as factory layout, setup time reduction and the operation of internal systems[58].

Supply chain development

Improving supply chain performance starts with mapping supply chains to understand the capabilities and functions of their members. Lean construction resists the temptation to reduce the cost or increase the speed of any single activity because this may reduce performance of the total chain. Improving wider system performance requires understanding (1) the source of materials and resources, (2) the processing and handling steps, (3) the use of time from order to delivery and (4) the value added along the way. Lead times and genuine 'drop-dead dates' for selecting design options can then be established[59].

Final comment

Many tools and techniques indicate the one-to-one management of the customer/supplier relationship. We are yet to see in construction, tools and techniques that reach through the supply chain.

Current research

It would appear that supply chain management could have far-reaching implications for the construction industry. The supply chains are complex – in addition to the relationships that exist between organisations that sell and buy products and services, numerous business transactions occur along the chain between customers and suppliers. The structural market forces that underpin the supply chains impact upon the nature of the competition within those markets and tend to impact directly upon the relationships between customers and suppliers. The nature of the competitive environments affects the sourcing strategies and procurement methods. Ultimately sourcing strategies and customer/supplier relationships have an impact upon project performance. There is a little doubt that there are varying degrees of strategic procurement and supplier management being undertaken between tiers in the construction supply chain. However, supply chain management has as an ideal that the entire chain from source to end user is managed in such a manner as to satisfy the end customer. If such a scenario were to be achieved in the construction industry it would require of clients to reach through more than one tier and proactively manage supply. This ideal has yet to be achieved.

References

1. New, S. (1997) The scope of supply chain management research. *Supply Chain Management*, **2** (1), 15–22.
2. Day, M. (1998) Supply chain management: business process effects in the ceramics industry. *Logistics Research Network 1998*. Cranfield School of Management, Cranbrook, UK.
3. Christopher, M. (1998) *Logistics and Supply Chain Management: Strategies for Reducing Costs and Improving Services*, 2nd edn. Pitman, London.
4. Poirier, C.C. & Reiter, S.E. (1996) *Supply Chain Optimisation: Building the Strongest Total Business Network*, Vol. 1, 1st edn. Berrett-Koehler, San Francisco.
5. Copacino, W. (1997) *Supply Chain Management: The Basics and Beyond*, Vol. 1, 1st edn. St Lucie Press, Boca Raton.
6. Ellram, L. (1991) Supply chain management: the industrial organisation perspective. *International Journal of Physical Distribution and Logistics Management*, **21** (1), 13–22.
7. Ross, D.F. (1998) *Competing Through Supply Chain Management*, Vol. 1, 1st edn. Chapman & Hall, New York.
8. London, K., Kenley, R. & Agapiou, A. (1998) Theoretical supply chain network modelling in the building industry. *Asso-*

ciation of Researchers in Construction Management (ARCOM) Annual Conference. ARCOM, Reading, UK.

9. Lambert, D.M., Pugh, M. & Cooper, J. (1998) Supply chain management. *The International Journal of Logistics Management*, **9** (2), 1–19.
10. London, K. & Kenley, R. (1999) Client's role in construction supply chains: a theoretical discussion. *Joint Triennial CIB Symposium W65 and W55*. Cape Town, South Africa.
11. DIST (1997) *Supply Chain Partnerships Program – Building More Profitable Inter-organisational Relationships*. ACT, Canberra.
12. Coyle, J., Bardi, E. & Langley, C. (1996) *The Management of Business Logistics*, 6th edn. West Publishing, Minneapolis.
13. London, K. & Kenley, R. (2001) An industrial organisation economic supply chain approach for the construction industry: a review. *Construction Management and Economics*, **19** (8), 777–88; http://www.tandf.co.uk/journals
14. Nishiguchi, T. (1987) *Competing Systems of Automotive Components Supply an Examination of the Japanese 'Clustered Control' Model and the 'Alps Structure'*. First Policy Forum, Tokyo.
15. Hines, P. (1994) *Creating World Class Suppliers*, Vol. 1, 1st edn. Pitman Publishing, London.
16. Harland, C.M. (1996) Supply chain management: relationships chains and network. *British Journal of Management*, **7** (Special), 63–80.
17. Porter, M.E. (1985) *Competitive Strategy: Creating and Sustaining Competitive Advantage*, Vol. 1. Free Press, New York.
18. Bowersox, D. & Closs, D. (1996) *Logistics Management: The Integrated Supply Chain Process*. McGraw Hill, New York.
19. Akintoye, A. (1995) Just in time application and implementation for building material management. *Construction Management and Economics*, **13**, 105–13.
20. O'Brien, W. (1997) Construction supply chains: case study, integrated cost and performance analysis. In: *Lean Construction* (ed. Alarcon, L.), Balkema, Rotterdam, 187–222.
21. Agapiou, A., Clausen, L.E., Flanagan, R., Norman, G. & Notman, D. (1998) The role of logistics in the materials flow control process. *Construction Management and Economics*, **16**, 131–7.
22. Tommelein, I. & Yi Li, E. (1999) Just-in-time concrete delivery:

mapping alternatives for vertical supply chain integration. *IGLC 7th Annual Conference*. Berkeley, California.

23. Graves, A. (2000) *Agile Construction Initiative*. University of Bath, Bath, UK.
24. Barlow, J. (1989) From craft production to mass customisation: customer focused approaches to housebuilding. *IGLC 6th Annual Conference*. Sao Paulo, Brazil.
25. London, K., Kenley, R. & Agapiou, A. (1998) The impact of construction industry structure on supply chain network modelling. In: *Institute of Logistics, Logistics Research Network*. Cranfield University, UK.
26. Womack, J.P., Jones, D.T. & Roos, D. (1990) *The Machine that Changed the World*, Vol. 1, 1st edn. Rawson Associates/Macmillan, New York.
27. Krafcik, J. (1988) Triumph of the lean production system. *Sloan Management Review*, **30** (1), 41–52.
28. Alarcon, L. (ed.) (1997) *Lean Construction: Compilation of 1993–1995 IGLC Proceedings*. Balkema.
29. Lamming, R. (1993) *Beyond Partnership: Strategies for Innovation and Lean Supply*. Prentice Hall, Hemel Hempstead.
30. Nishiguchi, T. (1994) *Strategic Industrial Sourcing: The Japanese Advantage*, Vol. 1, 1st edn. Oxford University Press, New York.
31. Sugimoto, Y. (1997) *An Introduction to Japanese Society*. Cambridge University Press, Melbourne.
32. London, K. (2001) The evolution of an alliance network to develop an innovative construction product: an instrumental case study. *CIB World Building Congress*. Wellington, NZ.
33. Green, S. (1999) The dark side of lean construction: exploitation and ideology. *IGLC 7th Annual Conference*. Berkeley, California.
34. Kamata, S. (1984) *Japan in the Passing Lane: An Insider's Account of Life in a Japanese Auto Factory*. Unwin Paperbacks, UK.
35. Ballard, G. & Howell, G. (1999) Bringing light to the dark side of lean construction: a response to Stuart Green. *6th Annual International Group for Lean Construction Conference*. Berkeley, California.
36. Aouad, G., *et al.* (1999) The development of a process map for the construction sector. *Joint Triennial Symposium CIB W65 and W55*. Cape Town, South Africa.
37. O'Brien, W.J. (1998) *Capacity Costing Approaches for Construction*

Supply Chain Management. Department of Civil and Environmental Engineering, Stanford University.
38. Lamming, R. & Cox, A. (1995) *Strategic Procurement Management in the 1990s: Concepts and Cases*. Earlsgate Press, London.
39. Gomes-Casseres, B. (1996) *The Alliance Revolution: The New Shape of Business Rivalry*. Harvard University Press, Cambridge, MA.
40. Langford, D.A. & Male, S. (1991) *Strategic Management in Construction*. Gower, Hants, UK.
41. Cox, A. & Townsend, M. (1998) *Strategic Procurement in Construction: Towards Better Practice in the Management of Construction Supply Chains*, Vol. 1, 1st edn. Thomas Telford Publishing, London.
42. Rowlinson, S. & McDermott, P. (eds) (1999) *Procurement Systems: A Guide to Best Practice in Construction*. E & FN Spon, London.
43. Olsson, F. (2000) *Supply chain management in the construction industry: opportunity or utopia*. Unpublished licentiate thesis, Department of Engineering Logistics, Lund.
44. Clausen, L. (1995) *Report 256: Building Logistics*. Danish Building Research Institute, Copenhagen, 4.
45. Vrihjhoef, R. & Koskela, L. (1999) Roles of supply chain management in construction. *IGLC 7th Annual Conference*. Berkeley, California.
46. Bancock, G., Baxter, R. & Davis, E. (1998) *Dictionary of Economics*. Penguin Books, London.
47. Hobbs, J. (1996) A transaction cost approach to supply chain management. *Supply Chain Management*, **1** (2), 15–27.
48. Slack, N. (1991) *The Manufacturing Advantage*. Mercury Business Books, London.
49. Piore, M. & Sabel, C. (1984) *The Second Industrial Divide – Possibilities for Prosperity*. Basic Books, New York.
50. Alter, C. & Hage, J. (1993) *Organisations Working Together*, Vol. 1. Sage Publications, London.
51. Eccles, R. (1981) The quasifirm in the construction industry. *Economic Behaviour and Organization*, **2**, 335–57.
52. Taylor, J. & Bjornsson, H. (1999) Construction supply chain improvements through internet pooled procurement. *IGLC 7th Annual Conference*. Berkeley, California.

53. London, K. & Kenley, R. (2000) *Mapping Construction Supply Chains: Widening the Traditional Perspective of the Industry*. European Association for Research Industrial Economics, Lausanne.
54. Hillebrandt, P.M. (1985) *Economic Theory and the Construction Industry*, Vol. 1, 2nd edn. Macmillan Press, London.
55. AEGIS (Australian Expert Group for Industry Studies) (1999) *Mapping the Building and Construction Product System in Australia.* Department of Industry, Science and Technology, Canberra.
56. Love, P. *et al.* (1999) Cooperative strategic learning alliances. *Joint Triennial CIB Symposium W65 and W55.* Cape Town, South Africa.
57. Riggs, D.A. & Robbins, S.L. (1998) *The Executives's Guide to Supply Chain Management Strategies: Building Supply Chain Thinking into all Business Processes*, Vol. 1, 1st edn. AMACOM, New York.
58. Hines, P. (1996) Network sourcing: a discussion of causality within the buyer-supplier relationship. *European Journal of Purchasing & Supply Management*, **2** (1), 7–20.
59. Ballard, G. & Howell, G. (2001) Lean Construction Institute website. http://cic.vtt.fi/lean/.

Chapter 8
Partnering and alliancing

Introduction

'Partnering' is difficult to define. It means many things to many people. Partnering has to do with human relationships, with stakeholders' interests, with the balance of power. In other words partnering has to do with human interaction and, as an inevitable consequence, it is a complex subject which is difficult to pin down and analyse. Partnering is more than simply formalising old-fashioned values, or a nostalgic return to the good old days when a 'gentleman's word was his bond', (although moral responsibility and fair dealing are an essential underpinning of any partnership[1,2]). It is more than a building-procurement technique (although building-procurement techniques can be used to operationalise good practice, bring about cultural change and thus create a more cohesive team[3]). The use of partnering in the construction industry has many advocates and many claims of success. The titles of journal articles on partnering positively exude confidence and self-assurance. Titles such as 'Partnering means making friends not foes[4]', 'Partnering pays off'[5], 'Partnering makes sense[6]' and more forcefully 'Partnering – the only approach for the 90s[7]', abound in the professional journals.

In this chapter we try to cut away the hype surrounding partnering, by tracing its origins, describing how it has been adopted by the construction industry and then describing the benefits and also the risks and pitfalls associated with the implementation of partnering.

The origins of partnering

The origins of partnering, as a construction management concept, are relatively recent, dating from the mid-1980s[8]. This not to say that partnering did not exist prior to that period and indeed many would subscribe to the view that 'Partnering between contractors and private clients is as old as construction itself'[9]. It has also been claimed that, in the UK, companies such as Bovis have developed a culture and tradition of non-adversarial relationships with particular clients since the 1930s[3].

For the purposes of this chapter we concentrate on the period from the mid-1980s onwards when then term 'partnering' was given quite explicit connotations. In effect we focus on *formal* partnering, where there is evidence of an explicit arrangement between the parties. This is not to dispute the existence and importance of *informal* partnering (or as it has been described 'partnering without partnering'[10]). However for the time being we discount informal partnering from our considerations.

According to the National Economic Development Office (NEDC) report *Partnering: contracting without conflict*[11], true partnerships in the formal sense only became established in the mid-1980s, the first being that between Shell and partners in 1984. The most frequently cited partnering arrangement of the 1980s is the Du Pont/ Fluor Daniel relationship for the Cape Fear Plant project. The partnering agreement between Du Pont and Fluor Daniel was made in 1986[11] and was a formalisation of a relationship which had existed since 1975. Other notable partnering relationships during this era were Union Carbide/Bechtel; Proctor & Gamble/Kellogg; and Shell Oil/Parsons.

Partnering in a construction industry context

Most commentators attribute the emergence of partnering as a force in the construction industry in the late 1980s to the work of the Construction Industry Institute of the United States (CII) and the adoption of partnering by the US Army Corps of Engineers (mainly through the efforts of Charles Cowan)[1,12,13]. In the present era, extensive examples of partnering can be found in the USA with the movement gaining momentum in New Zealand, Australia and the UK. In the latter two countries this gain in momentum is partly as a result of prompting by the Gyles Royal Commission into Productivity in the Building Industry in New South Wales[13] and the Latham Report in the UK. Latham, in his foreword to *Trusting the team: the best practice guide to partnering in construction*[14], states that 'partnering can change attitudes and improve the performance of the UK construction industry. I hope that the industry and its clients will now use the report to embark upon partnering.'

In Australia the Gyles Royal Commission went further than this and carried out a pilot study on partnering as a means of encouraging a cultural shift in the New South Wales construction industry[13]. Also as we mentioned in Chapter 1, partnering is one of the approved criteria in the New South Wales Department of Public Works and Services contractor accreditation scheme[15], demonstrating the importance which some government clients place on the introduction of partnering.

The extent of the adoption of partnering by the construction industry at large is still difficult to quantify. There are however numerous examples from the USA of successful partnering. A 1994 study[16] of 2400 attorneys, design professionals and contractors rated project partnering and mediation as top of the list of alternative dispute resolution methods. The 1994 annual meeting of the Construction Industry Institute in Boston reported that

> 'in terms of programs such as safety, constructability, total quality management and both long-term and short-term partnering that long-term partnering offered the most impressive savings. CII's strategic alliance task force reported that 196 projects using long-term partnering saved an average of 15% of total installed cost. On project-specific partnering, CII's team-building task force recorded an average of 7% savings on five very large projects'[17].

These statistics indicate the general level of the acceptance of partnering in the USA.

In Australia partnering is relatively commonplace to the extent that the Master Builders Association runs an annual competition which attracts a range of entries of good examples of partnered projects. Hellard[1] is somewhat sceptical of the extent of the uptake of partnering in the UK.

> 'A full partnering project embracing client, design team, general contractor, and subcontractors with a formal project charter and formal training and commitment sessions does not yet seem to have emerged.'

However several case studies of UK partnering are included in the recently published *Trusting the Team: the Best Practice Guide to Partnering in Construction*[14].

The goals of partnering

We said at the outset of this chapter that partnering is a difficult phenomenon to isolate and define. However defining the *goals* of partnering, as opposed to defining the *nature* of partnering, is more simple. There are a number of definitions in circulation on the goals of partnering. Some of these are very broad, for example 'partnering is a process for improving relationships among those involved on a construction project to the benefit of all'[18]. Others are much more detailed but share the same philosophy, for example:

> '[Partnering] is not a contract but a recognition that every contract includes a covenant of good faith. Partnering attempts to establish working relationships among stakeholders through a mutually developed formal strategy of commitment and communication. It attempts to create an environment where trust and teamwork prevent disputes, foster a co-operative bond to everyone's benefit and facilitate the completion of a successful project[19].'

All commentators stress the achievement of trust and co-operation as the essential goal of partnering. Typically this is described as 'a long term contractual commitment between two or more organisations based on a spirit of trust and co-operation. The idea is to allow each participant to make the most of his resources and continually improve performance'[20].

Cowan, one of the principal architects of the modern partnering movement, stresses that:

> 'Partnering is more than a set of goals and procedures; it is a state of mind, a philosophy. Partnering represents a commitment of respect, trust, co-operation, and excellence for all stakeholders in both partners' organisations[12].'

Categories of partnering

There are two different categories of partnering and within these categories there are a variety of types. The two categories of partnering are strategic partnering and project partnering. (Strategic partnering is sometimes referred to as 'multi-project partnering' or less frequently as 'second-level partnering' and project partnering as 'single project partnering' or 'first-level partnering'.) 'Strategic partnering takes place when two or more firms use partnering on a long-term basis to undertake more than one construction project'[14]. Project partnering is the converse of this, and occurs when two or more firms come together in a partnering arrangement for a *single* project. In the USA, 90% of all partnering is project partnering[14]. Because partnering has to do with long-term relationships, it follows that more gains and benefits are likely to be achieved from the longer-term strategic partnering as opposed to the shorter duration project partnering arrangements. However project partnering can be a stepping stone to strategic partnering.
[Note: For the purposes of this text we use the term 'project' to mean a construction industry project of a 'normal' timescale. Previously we cited the Du Pont/Fluor Daniel, Cape Fear Plant project as an early partnering agreement. Although this was a single project, it was of such massive dimensions, both in timescale and physical size, that this would most properly be categorised as strategic rather than project partnering. This however is the exception that proves the rule. In most cases the differentiation between 'project' and 'strategic' is relatively straightforward.]

Project partnering

The participants

The first issue to be to be addressed in any form of partnering is: who should participate in the partnering arrangement? This can immediately give rise to a 'chicken and egg' syndrome. If the partnering concept is introduced at the outset of the project, this will have a bearing on the procurement method adopted which in turn will have a bearing on the composition of the project participants. In order to gain the maximum benefits of partnering, the general trend is to use a design and build (design and construct in Australia) form of procurement. The advantage of this approach is that it allows all the key stakeholders – i.e. the client and the design and build contractor (whose organisation includes the design team) – to be involved in a partnership arrangement from the outset of the project.

Partnering is not however the exclusive domain of design and build and can be used in a traditionally procured project where the lowest bid contractor is brought on board at the tender acceptance stage. In this situation the contractor is excluded from the design stages. Thus during the preconstruction phase the partnering arrangements can only take place between the client and the design consultants, the contractor being brought into the partnering relationship at the downstream stage of tender acceptance. This is not the ideal environment for partnering and has been likened to bringing a new player into a bottom of the league football team halfway through the season and then expecting them to become league champions[10].

The selection of the members of the partnering team is clearly of paramount importance in a concept which is all about trust and mutual support. The partnering team must be capable of carrying out its responsibilities throughout the project. 'If any one of the team members is not capable of carrying out its duties, any attempt at partnering will fail'[10]. However the task of selecting partners is not an easy one. As Cowan[12] remarks: 'Ideally, you want to select contractors or owners who have established a successful track record of partnering on previous contracts. By interviewing contractors an owner can discern interest and/or expertise.' At first glance this statement may seem to be a tautology. However it is possible for a contractor to progress from being an interested party to a partnering participant. This progression is described in the New South Wales Contractor Accreditation Scheme[15] which identifies five categories of partnering involvement ranging from low to high (Table 8.1). These categories are as follows:

Table 8.1 Levels of partnering characteristics[15].

Level	Stage	Characteristics
Low	Recognition stage	Aware of benefits of partnering
Average	Development stage	Developing partnering policy
Above average	Establishment stage	Committed to use of partnering in all projects
Good	Continuous improvement stage	Documented evidence of ongoing improvement in project delivery through partnering
High	Best practice stage	Record of long-term relationships with consultants, suppliers and subcontractors

Commitment

The decision to partner is a very significant one because of the reliance which one partner must inevitably place on the other/s. It follows that a decision to partner should be taken at the highest possible level in an organisation. 'It is inconceivable that … partnering could be carried out without commitment at the very highest levels within both organisations'[11].

Although great stress is made by all commentators on the need for high-level commitment there is also a need for 'internal partnering' to take place within an organisation prior to a commitment by senior management to enter into a partnering agreement.

> 'Internal partnering means preparing your organisation and reviewing internal procedures and documentation. It also involves educating and informing staff on partnering. The result of this process should be a united organisation that is prepared to work closely with another organisation'[14].

The partnering process

It is inevitable that the partnering process will have many forms and many variations. In this section we consider single project partnering, the processes for which will be different to strategic partnering which we consider later. The process which is adopted for project partnering will depend on the circumstances of each situation. If the project participants have previously been involved in similar partnering arrangements, many of the bridges will already have been made, whereas if this is a first-off occasion for the participants then the process will have to be developed with particular care and sensitivity. For the purposes of this section we assume that we are dealing with partnering with first-time partners.

Fundamentally the partnering process is about team-building which is why the function of internal partnering is so important in achieving a successful outcome. The processes of partnering must be seen as a means to an end and must not be seen as an end in itself. Katzenbach and Smith state that

> 'a demanding performance challenge tends to create a team. The hunger for performance is far more important to team success than team building exercises, special incentives, or team leaders with ideal profiles'[21].

It is essential that the team are enthused by the challenge of partnering, and the generation and maintenance of this enthusiasm is one of the primary functions of the partnering process[1].

In general the partnering process falls into three phases: the preproject stage, the implementation stage and the completion or feedback stage.

Preproject stage

The preproject stage begins with the decision of whether or not to partner. In some cases a readiness to use partnering will be a condition of prequalification for government tenders[15]. Whether or not the initiative to partner comes from the client, design team or contractor, it is a potentially high-risk decision and as such cannot be taken lightly. It is not an automatic assumption that partnering is always the best approach. For some parties and in some situations partnering should be best avoided. Ideally there should be a synergistic relationship between the parties with each party bringing complementary strengths to the partnering table. The worst case scenario is where an inherent weakness in one party spreads throughout the team: in this situation partnering is likely to exacerbate rather than stabilise the situation.

Assuming that the decision to partner is taken, how then is the process of selection initiated? There is no clear answer to this question. In some situations, such as the New South Wales Department of Public Works and Services (DPWS), partnering may be a direct requirement of prequalification to tender for government projects. The partnering connections must however be made by the individual parties, albeit with the encouragement of DPWS. In other instances the initiative to partner may come from an individual within a company. Award schemes such as the Australian Master Builders partnering awards help to sow the seed, as do publications such as *Trusting the Team: the Best Practice Guide to Partnering in Construction*'[14]. Given that partnering has to do with trust and mutual co-operation, the process of making the initial overtures between potential partners is never going to be prescriptive and to some extent it will always be shrouded in mystery and cloaked with commercial confidentiality. The following overture from a client to a design and build contractor gives a flavour of how a partnering arrangement could be initiated:

> 'In order to accomplish this contract most effectively, the Owner proposes to form a cohesive partnership with the Contractor and its subcontractors. This partnership would strive to draw on the strengths of each organisation in an effort to achieve a quality product done right the first time, within a budget and on schedule. This partnership would be bilateral in make-up and participation will be totally voluntary. Any cost associated with effectuating this partnership will be agreed to by both parties and will be shared equally with no change in the contract price[22].'

As can be seen this is a voluntary agreement which is aimed at creating a culture of mutual trust as opposed to an adversarial climate. (The reader may feel at this stage that the conventional construction contracts are, of their very nature, adversarial in approach and that this presents something of a dichotomy between the goals of partnering and the normal contractual agreements. This is a difficult issue and will be dealt with later in the chapter under contractual and legal issues. For the moment we leave these qualms aside.)

Having achieved an understanding to partner between top management, the progression of the partnering process takes a more predictable path. Figure 8.1 is a stylised flow chart of the partnering process.

Initial partnering workshop

Once the decision to partner has been made, the stage is set to bring together the key middle managers who will be involved in the project on a day-to-day basis. In a design and build procurement approach this group will comprise a range of stakeholders from the client body, the design team and the contractor. There may also be legal and financial advisers and on some occasions even representatives from statutory planning authorities and building control. In the interests of group dynamics the size of the group should not be overly large and should be restricted to a maximum of about 25[14]. The team is normally introduced to one another through an initial partnering workshop (the detailed mechanisms applied at this workshop will vary according to specific circumstances and also according to the culture of different countries). The purpose of this workshop is to agree the ground rules for the way ahead and should be held as soon as possible after the contract has been awarded. There seems to be general agreement that this workshop should be held on neutral territory, preferably at a 'retreat', away from all other distractions, and would normally be of one, two or three days duration depending on the nature of the project and the familiarity of the participants with partnering. It is highly recommended that the workshop sessions are managed, for at least part of the time, by a facilitator with experience of partnering.

The goals of the workshop are:

> 'to open communications, develop a team spirit, establish partnering goals, develop a plan to achieve them, and gain commitment to the plan[22]'.

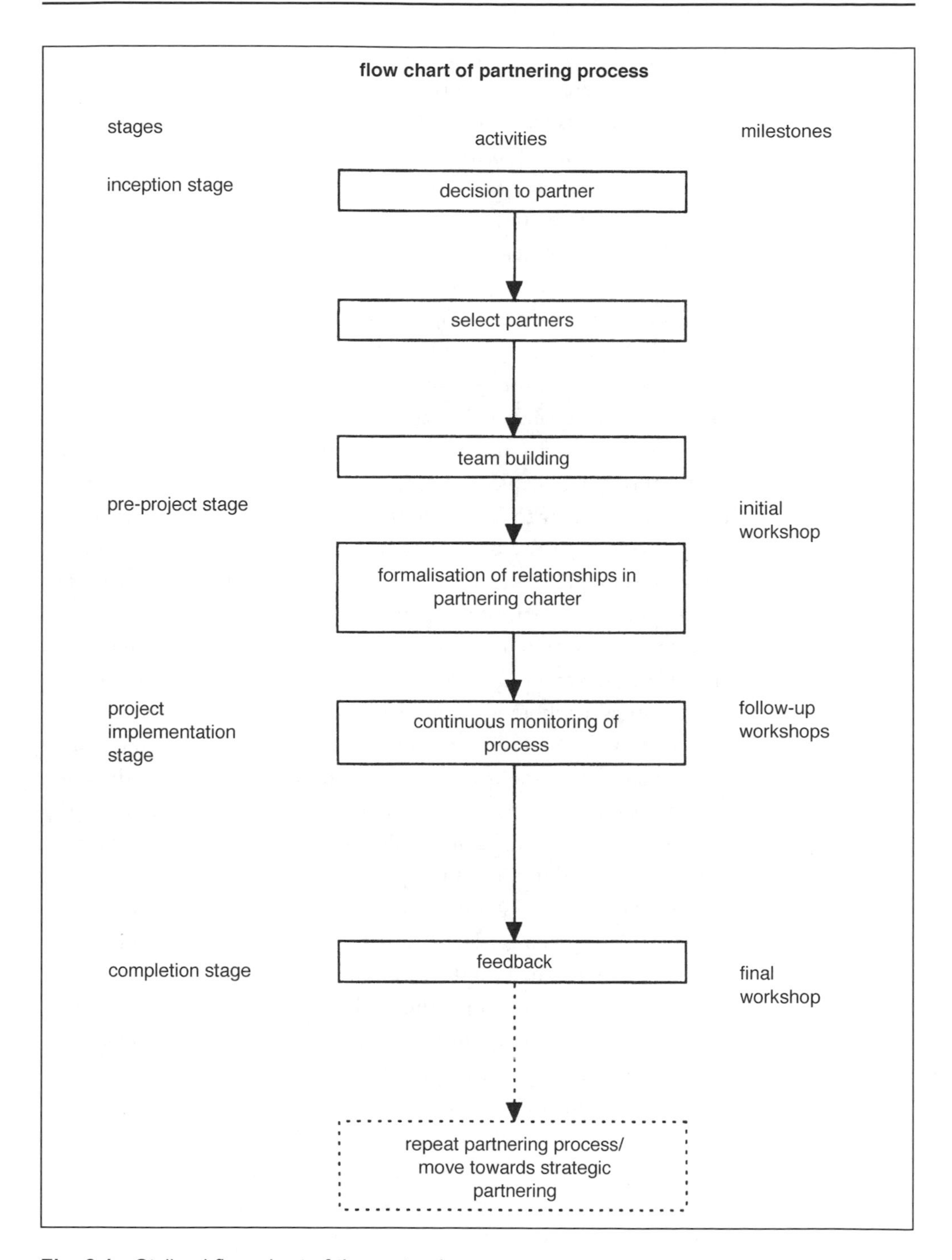

Fig. 8.1 Stylised flow chart of the partnering process.

The general pattern of the workshops is to begin with team-building exercises, where participants are encouraged to explore mutual interests of a personal nature. This may be done in quite a low-key way, or may be more formalised by using devices such as the Myers-Briggs Type indicator as a means of personality profiling. The team then moves on to develop a conceptual framework or common mental picture of what partnering is about and how this will influence their own participation in the project. The team then goes on to consider and agree mutual objectives and mutual interests on the lines of a SWOT analysis. (SWOT analysis means identifying Strengths, Weaknesses, Opportunities and Threats.) Mutual objectives such as time, cost and quality will feature large on the agenda with other issues such as safety, constructability, and electronic data interchange (EDI) also being explored. The mechanism for arriving at these objectives is usually to form action teams from within the participants to deal with the many issues which will emerge during the course of the workshop. These action teams may comprise the counterparts of the disciplines of each of the partners, for example legal advisers, or may contain a mix of disciplines such as architects and engineers. In any event the action teams will be tasked with bringing solutions to the group as a whole on issues such as deriving performance indicators, dispute resolution or minimisation and safety issues. The action teams assist the group-bonding process and reduce the partnering issues to manageable dimensions. In many instances the members of the action teams will remain in close communication throughout the duration of the project.

The first workshop session culminates in agreement of a partnering charter and in agreeing a partnering implementation plan. Agreement should also be reached on the timing of follow-up workshops which are essential in maintaining the will to succeed in the partnering arrangements. The action teams will be tasked with the development of their assigned topics for the follow-up workshop, which would typically be held at three monthly intervals over the duration of the project.

A partnering charter (or project charter) will always be signed by the participants. (The signing of the charter by all the participants is seen as a very symbolic act and is a tangible way of swearing an oath of allegiance to one another and to the recognition that individual interests are subordinate to those of the project.) The signatories to the charter (in a design and build scenario) could come from the following organisations: client, contractor, design team, sub-contractors, manufacturers, suppliers, consultants (such as legal advisers and planning experts).

The actual charter may be quite general or quite detailed depending on the preferences of the partnering team and the nature of the project. Given that copies of the charter will be posted on the walls of all the participants and be displayed in a variety of locations from the client's office to site workshops and site offices, then a single A4 sized sheet is the limiting factor. An example of a partnering charter is shown in figure 8.2.

Partnering Charter

We the Partners of XXXXX Project are committed to working together in a trusting and sharing environment, and are dedicated to achieving excellence for the benefit of all the stakeholders involved in this project.

The goals of this project are to:

Complete the project on time
Complete the project within budget
Achieve a reasonable profit margin
Have zero lost time due to accidents
Produce a high quality building with zero defects

Our team mission statement is that:
We will work in an environment of open door communications committed to achieving excellence in the XXXXX Project.

Signed by:

_______________ _______________
_______________ _______________
_______________ _______________
_______________ _______________
_______________ _______________

Fig. 8.2 Partnering Charter.

To summarise, irrespective of the mechanism used to forge a partnering agreement,

> 'before the project starts, participants need to establish a common charter for the project, agree on performance criteria and how they will be measured, develop mechanisms for resolving conflict, and establish general guidelines for working together. Considerable time and energy must be invested up-front to establish the foundation for a working relationship that can tolerate conflicts and avoid costly misunderstandings'[12].

Implementation stage

At the conclusion of the initial partnering workshop an implementation plan will have been established which will include a timetable for a series of follow-up workshops. The frequency and the duration of these workshops will depend on individual circumstances. The initial workshop is geared towards changing a traditional adversarial culture to a team-spirited win-win environment. This quantum leap in human relations needs constant nurturing (particularly with newcomers to partnering). Follow-up workshop sessions are essential in reinforcing the partnering culture. The follow-up workshop sessions usually follow a similar pattern to the initial workshop. If an external facilitator has been used for the initial partnering workshop, it makes sense to use the same facilitator for the follow-up sessions and, as with the initial workshop, it is recommended that the follow-up sessions are held on neutral territory away from distractions.

It is important that the parties to the Partnering Charter agree to evaluate the team's project performance against an agreed set of criteria. Cowan *et al.*[12] suggest a simple but formal scaled questionnaire which, in addition to criteria such as cost control, safety record and time scheduling, also includes criteria which are explicit to the partnering relationships such as teamwork and problem-solving. The Royal Commission into Productivity in the Building Industry in New South Wales, which introduced the partnering concept to Australia, used a ten-point partnering effectiveness monitor in its pilot study in 1992.

A typical partnering evaluation summary, or partnering effectiveness monitor, is shown in figure 8.3.

For projects of more than one year duration, formal written partnering evaluations might take place every 1–3 months with follow-up team meetings every 3–6 months. These formal evaluations will be a key item on the agenda of the follow-up workshops.

In addition to the scheduled follow-up partnering workshops,

Partnering evaluation summary

No.	Objective	Last period			This period			
		Weight	Rating	Score	Weight	Rating	Score	Comments
1	Quality achieved	15	4	60	15	4	60	
2	Cost controlled	15	4	60	15	5	75	
3	Time performance	15	3	45	15	4	60	
4	Teamwork	10	4	40	10	4	40	
5	Safety	10	3	30	10	3	30	
6	Avoidance of industrial disputes	10	4	40	10	2	20	
7	Avoidance of litigation	10	4	40	10	4	40	
8	Satisfactory cash flow	5	5	25	5	4	20	
9	Environmental impact contained	5	3	15	5	3	15	
10	High morale and job satisfaction	5	4	20	5	5	25	
Overall satisfaction of partnership stakeholders		100	1 to 5	375	100	1 to 5	385	

Fig. 8.3 Partnership evaluation summary.

informal partnering sessions will also take place. Although these informal sessions are important in terms of maintaining good relationships between the partners, it is equally important that if and when new members join the team, they are inducted into the partnering process via a formal workshop session rather than through an informal session. This is because there is always a danger that the ethos of partnering will become diluted (particularly with participants who are new to partnering) in the dissemination process.

Partnering pitfalls during the implementation stage

Project partnering is a high-risk, high-gain approach. Naturally many of the problems which can be identified during the implementation stage of the partnering process also apply to non-partnered conventionally procured projects. However although there is a greater potential for stakeholders (those involved in the partnering process) to gain from partnering compared to the conventional process, conversely, there is also a greater risk of loss to the stakeholders if things go wrong.

Kubal[10] lists six criteria as being essential to successful partnering. It could be argued that these criteria are equally essential to the success of a non-partnered process. However if these criteria, which are summarised below, are not in place, the partnered process is at greater risk.

- There must be no weak links in the team members. They must all be capable of performing their tasks. (Bennett and Jayes[14] go as far as to suggest that team members be allowed two 'mistakes' after which consideration should be given to their being replaced.)
- Owners must be properly represented and active in the project.
- The composition of the partnering team should, wherever possible, remain stable for the duration of the project.
- Partnering should start at the design stage.
- There is a need to appoint a team leader or 'champion' to ensure that partnering principles do not slip out of focus.

❑ The involvement of major subcontractors and manufacturers is crucial, as is the integration of trade contractors.

These six criteria underscore the vulnerability of partnering relationships and highlight the challenges involved in implementing a successful partnering programme.

> 'The fragility of interorganisational alliances stems from a set of common "dealbusters" – vulnerabilities that threaten the relationship. Partnerships are dynamic entities, even more so than single corporations, because of the complexities of the interests in forming them. A partnership evolves; its parameters are never completely clear at first, nor do partners want to commit fully until trust has been established. And trust takes time to develop. It is only as events unfold that partners become aware of all the ramifications and implications of their involvement[23].'

There is of course no concrete evidence that partnering will always benefit the project. Although on the one hand partnering can be viewed as 'saving money and making work more enjoyable' there are also disadvantages[14]. Cost is one example. The direct costs of partnering are the cost of the workshop sessions and also the cost of the partnering facilitator, hiring of venues, transport, etc. The cost of the participants' time in attending workshops, etc. is not an additional direct cost of partnering but is a redistribution of costs which would normally occur, or at least should occur in the traditional procurement process. In addition other potential disadvantages and indirect costs can be identified as follows[14]:

❑ **Stale ideas** – due to the lack of stimulation which can occur when the same partners are in a stable relationship and the stimulus of new players is missing.
❑ **The cost of consultants may be higher** – the partnering process produces more alternative solutions. As a result design consultants have a greater workload which should be reflected in higher fees.
❑ **Reduced career prospects** – staff involved in partnering may see this as an additional burden in terms of workload, which will not count towards their career advancement.

- **Loss of confidentiality** – there can be problems associated with this unless proper safeguards are taken.
- **Investment risk** – investment in joint development for the project, for example in developing an EDI (electronic data interchange) system, may be risky if the single project partnership does not extend into future contracts.
- **Dependency risk** – there is a risk that the partners become too dependent on each other. They should therefore spread their partnering agreements over a range of partners. On occasion however, short-term financial aid may need to be given to a single source supplier in order to ensure the supplier's continued existence.
- **Corruption** – the one-off relationship of non-partnered projects minimises the possibility and opportunity for corruption. There is an additional obligation on partnerships which may become long-term in nature to ensure that safeguards and checks against corruption do not become lax.

The underlying cautionary note which links all of these potential disadvantages of partnering is that, to be successful, partnering needs an equal level of commitment from all the partners.

> 'Unevenness of commitment often develops from the basic differences between organisations. If for example, a small contractor is entering into a public works or defence contract managed by a large government agency, the contractor may feel that it is not possible to devote the time and staff to partnering that will make the contractor an equal partner. Every effort is needed by all involved parties to balance the commitment on both sides[22].'

If this commitment is given, many of the pitfalls of implementing partnering will be avoided.

Limits to partnering

There is general agreement among commentators that participants in partnering should not put all their eggs in the one basket, and should limit the extent of their partnering commitments. As early as 1981 the UK National Economic Development Office[11] recommended that

> 'In general it was considered that no single partnering arrangement should utilise more than 30 per cent of the resources of the office in which it was located, and that a contractor's total commitment to partnering should not utilise more than 50 per cent of total technical and managerial resources.'

Completion stage

The completion stage of the partnering process is particularly important, and is one of the particular strengths which the partnering process has in comparison to the traditional project team approach. In this latter type of arrangement there is rarely a formal procedure that attempts to encapsulate the knowledge base which has accumulated as a consequence of the project.

The completion stage of the project partnering process would normally be centred round a final workshop session. Usually this final workshop will incorporate a lunch or dinner at which senior management will take the opportunity to make mention of specific contributions and highlight the characteristics of the completed project. Although this social event is an occasion for emotional release and for self-congratulation (particularly if the project has gone well), the final workshop is not simply to do with social niceties. It is an opportunity to undertake some serious work in order to learn from the completed project. Whereas the initial workshop and follow-up workshops had to do with engendering team spirit and improving the performance of the project *in-hand*, the final workshop is aimed at consolidating the ground for continued partnering relationships between the various team members. It is also the opportunity to return to the SWOT analysis of the initial workshop. The contributors to the final workshop should analyse objectively what went well and what was less successful in terms of both the project outcome and the partnering process. If the formal evaluation procedures (see implementation stage) have been assiduously carried out for the duration of the project, the progression of the project will have been fully charted (warts and all), and this will provide a good objective focus for the final workshop.

The participants at the final workshop should ideally comprise the participants from the first workshop plus the key stakeholders who joined the partnership team during the progression of the project, provided that the number of participants for the final workshop does not become too unwieldy. It has to be borne in mind that the final workshop is an important occasion for concentrated work, and may prove to be the stepping stone from single project partnering to strategic partnering. To this end, it is clearly of particular importance that the client body be as fully represented at the final workshop as at the initial workshop.

Strategic or multi-project partnering

There seems to be quite clear evidence that strategic, or multi-project, partnering will produce more substantial benefits than single project partnering[14,17]. The difference between single project partnering and strategic partnering is that strategic partnering has the added dimension of 'the development of a broader framework focusing on long term issues[14]'.

Strategic or multi-project partnering occurs where the partnering team is in a position to enter into an undertaking for a series of projects, either new-build or a rolling maintenance agreement and, as with single project partnering, can operate within conventional procurement agreements. Although there are significant benefits to be gained from strategic partnering, there are also significant hurdles to be overcome in its implementation, particularly with respect to national and international laws relating to free trade. For example in the USA, partly as a result of the anti-trust laws, over 90% of all partnering projects are of a single project nature[17] and, while the Central Unit on Procurement of the UK Treasury predicts a significant increase in partnering in the next few years, there is an implied caveat in its statement that

> 'Government policy is to use competition to achieve best value for money and to improve the competitiveness of its suppliers. *But* (our italics) the Government also appreciates the benefits of part-

nering and, in appropriate circumstances, these arrangements are being actively promoted[14].'

Despite these free trade reservations on partnering in general and strategic partnering in particular, it could be said that strategic partnering has all the advantages of single project partnering, only more so. The longer term nature of strategic partnering allows, among other things, for the development of a mutually beneficial physical infrastructure, for example, shared office accommodation and electronic data interchange (EDI). Although it is quite possible to create these arrangements for a single project, in highly technical and high-cost developments, such as EDI, there are clear financial and technical benefits to be had in the semi-permanent relationship of strategic partnering.

It would appear that the attendant risks with strategic partnering are no higher than those associated with single project partnering, although in many cases, but not always, the magnitude of the turnover and workload will be significantly higher in strategic partnering. It is worth noting that for any partnership to work, the partners must bring complementary skills to the partnership. There is no point in the partners bringing the same or similar strengths, nor should one partner be motivated by the notion of eventual takeover of the other partner. These points are valid for both forms of partnership but are particularly pertinent to a strategic partnering relationship.

Most strategic partnerships will come about through experience gained on single project partnerships. It follows therefore that the stakeholders in a strategic partnering arrangement are likely to be experienced in the mechanics of partnering. It is highly probable that the strategic partners will already have been involved with one another on individual projects. Strategic partnering operates at the macro level and single project partnering at the micro level. Kubal[10] uses the term 'second-level partnering' to describe long-term or strategic partnering, demonstrating second-level partnering by drawing on examples from the manufacturing sector. Here, established partnering arrangements between manufacturers and suppliers employ the 'just-in-time' inventory systems resulting in the supplier ensuring a steady workload, the manufacturer reducing production costs and the customer receiving goods at lower cost and higher quality.

While the goals and objectives of strategic partnering are empathetic with the goals of project partnering, they are not the same. Although terms such as time, cost, quality service and value can equally apply to project and strategic partnering, the context and timescale in which they are applied is different. Similarly, although for example, the mechanisms for strategic partnering appear to be similar to those of single project partnering, with the use of initial and follow-up workshops using an external facilitator, the timetable for these workshops may be quite different, with a period of say six months elapsing from partner selection to the start of the first project of the strategic partnering relationship. Moreover 6 to 18 months may elapse from the time that management decides to explore strategic partnering until the commencement of the first project[14].

The aims and objectives of strategic partnering are encapsulated in the signing of a partnering charter, on similar lines to the single project partnering charter. However whereas the signatories to the project charter may number some 25 or so participants, the signatories to a strategic partnering charter will be restricted to one per partner. The charter will reflect the values, beliefs, philosophy and culture of the partners and these attributes will normally be translated into a charter in the form of a mission statement followed by a series of objectives.

A typical strategic partnering charter would be as illustrated in figure 8.4.

As can be seen from the partnering charter, its aims and objectives are generic rather than project-specific. Some charters may include a statement to the effect that 'nothing in this statement constitutes a partnership or binding agreement'[14] to emphasise the non-legal nature of the agreement. Given that this charter will be displayed in a variety of office and site locations, it may be felt that stressing the non-legally binding nature of the charter is a redundant statement which detracts from the spirit of trust expressed in the partnering charter.

Legal and contractual implications of partnering

[Note: The legal and contractual implications of partnering will vary from country to country depending on the legal system in place. The following comments are made mainly in the context of the UK,

Strategic Partnering Charter

Mission statement

Our mission is to work towards continually improving the spirit of trust and business relationships between our organisations so that we produce a better quality environment for our end users.

Objectives

To ensure value for money for our buildings users

To maintain a spirit of trust in all our relationships and co-operative activities

To demonstrate a genuine concern and interest in the productivity of all stakeholders

To produce buildings of excellence

To produce our buildings within time, on budget with zero defects

To ensure a reasonable profit for all the stakeholders

Company A...........................

Company B...........................

Company C...........................

Company D...........................

Fig. 8.4 Strategic Partnering Charter.

Australia and New Zealand where the legal systems have a common basis, although even between these three countries there will be significant differences in terms of the trade practices in operation in each country. The added dimension of European Competition Law also affects the UK.]

Making explicit statements that a partnering charter does not create a legally binding relationship between partners does not necessarily mean that none exists. Certainly the strength of the relationship is much less than in a true partnership. In this latter case each partner is considered to be the agent of the partnership and, therefore, can bind the partnership with respect to third parties (a fiduciary duty)[14]. However as the Australian Construction Industry Institute report on partnering observes:

> 'Although the construction contract provides a framework of rights and obligations, partnering has the potential to impact upon the allocation of risk established by that contract and subsidiary contracts. If the partnering arrangement breaks down, a party may find itself in a position where it is necessary, or at least attractive, to assert that the contractual risk allocation has been altered, either by the provisions of the partnering charter or by subsequent conduct or representations in the course of the partnering process. This is, potentially, the major risk to partnering in Australia[24].'

It is not our intention in this text to explore the legal ramifications of partnering, particularly since these will vary from country to country. However it is worth stressing that partnering relationships whether strategic or single project will have, if not ramifications, at least implications in any country in which it is practised. The following example, extracted from the CIIA report[24], illustrates this by listing the ways in which the partnering process could impact on the underlying construction contract:

- ❑ The implication of a contractual duty of good faith
- ❑ The creation of fiduciary obligations
- ❑ Misleading and deceptive conduct
- ❑ Promissory estoppel and waiver
- ❑ Confidentiality and 'without prejudice' discussions.

With respect to 'good faith', the CIIA report makes the following observations.

> 'Although Australian courts have, as yet, not been prepared to uphold a duty to perform a contract in good faith there is an indication that given a breakdown in the partnering arrangement ... it would be open to a party to argue, that as the fundamental characteristics of partnering are consonant with good faith, and it is a US concept supported by the duty of good faith applied by the

> courts in that jurisdiction, there should be implied into its contractual relationships a general duty to act in good faith. Thus partnering could give rise to a situation where the court is prepared to recognise the concept of good faith; and therefore a party's rights to exercise its strict contractual powers might be limited by the duty to act in good faith.'

It should be noted that good faith is not the same as the duty to act reasonably. It is a broader subjective concept. Reasonableness, in legal terms, is a narrower objective standard. The report goes on to say that

> 'Finally, it may be open to a party to argue that the partnering charter forms part of the contractual documentation, even if there is no reference in the construction contract to partnering, and therefore, that the charter provided for the express duty to act in good faith in the performance of the contract.'

This legal opinion is given by way of illustrating the potential contractual and legal risks of partnering. (For a more detailed discussion of this aspect of partnering readers are referred to the CIIA report *Partnering: Models for Success. Research Report 8*, with the caveat that this report refers specifically to partnering in an Australian context.[24]) We are not inferring that partnering is a legal minefield. However it is sensible to attempt to minimise any attendant risks, both legal and contractual, in a partnering agreement by taking precautions and anticipating potential legal and contractual issues during the initiation stage of the partnering process. CIIA recommend that the parties to the agreement should incorporate provisions and procedures into the partnering charter as follows:

- **Good faith** – by express provision clarify the issue for the partnering arrangement so that it will not be implied into the contractual arrangement.
- **Fiduciary obligations** – (an obligation to act to a higher standard of conduct than a simple commercial relationship, a feature of a true partnership). By express provision exclude fiduciary obligations arising, thus permitting the parties to freely pursue their own interests; or alternatively, limit the scope of the

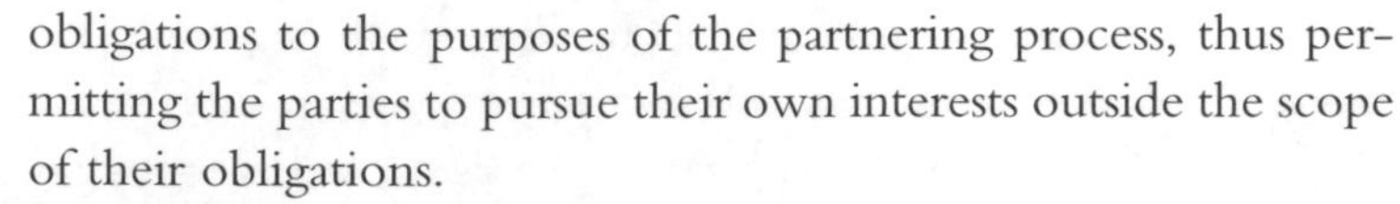

obligations to the purposes of the partnering process, thus permitting the parties to pursue their own interests outside the scope of their obligations.

- **Trades Practices Act** – it is very difficult to disclaim liability for misleading conduct by reference to a documented agreement.
- **Promissory estoppel and waiver** – (concerned with enforcing representations or promises as to future conduct, including promises not to rely on a party's strict legal rights, in circumstances where it would be unconscionable to do so). Incorporate into the partnering charter a procedure which must be followed if a party is to be denied its right to insist on enforcement in accordance with the construction contract's terms and conditions.
- **Confidentiality and 'without prejudice' discussions in relation to confidentiality** – include a confidentiality clause protecting confidential information by prohibiting it from being disclosed to parties other than the participants and for purposes other than the partnering process. In relation to the 'without prejudice' discussion, provide that any disclosure or concession made during the partnering process is made for that purpose, including any formal proceedings.

The CIIA report recommends that these provisions be incorporated into the partnering charter. It would appear to be difficult to incorporate these provisions into the partnering charter itself given the need for the charter to be an unambiguous statement of trust and co-operation which will be displayed in a variety of locations. It would be anathema to the partnering process to have a set of legal qualifiers in small print in the partnering charter. However the CIIA report gives sound advice which should be heeded. The solution is perhaps to document the types of issues listed in the CIIA report, not on the partnering charter itself, but as supporting documentation. The advice given in the CIIA report is not a return to the adversarial climate of the traditional approach, but a sensible precaution to minimise the legal and contractual risks to the participants.

Those undertaking partnering arrangements in the UK must also be aware of European law as it affects trading within the European Union. *Trusting the Team: Best Practice Guide to Partnering in Construction*[14] gives a detailed account of contractual and legal issues as they affect the UK. Those readers with particular interests in UK partnering are referred to this report. The following statement extracted from the *Trusting the Team* report clearly illustrates the importance of being attuned to contextual, contractual and legal issues:

> 'Partnering will always be unlawful in EU law if its effect is to discriminate against undertakings on national grounds, or breach the fundamental freedoms of the European Union – namely the freedom of movement of goods, services, workers and capital. Even if a partnering arrangement does not affect trade between member states, it may still be void, punishable by fine or liable in damages under English law if the agreement imposes restrictions on the price and supply of goods.'

Dispute resolution

It would be something of a utopian ideal to assume that disputes will not arise, even in the non-adversarial culture of a partnering relationship and, as with contractual and legal issues, it is prudent to have contingencies in place should the eventuality arise. In a survey of partnered projects in Australia[24], 91% of respondents agreed that a dispute resolution plan was an essential element of a partnering arrangement. In projects where partnering arrangements had been unsuccessful, 43% of these had no formal dispute resolution plan in place.

Although it is important to have dispute resolution procedures in place, it must be acknowledged that the incidence of disputes is likely to be less in a partnering relationship, because the partnering implementation plan should incorporate mechanisms preventing escalation of differences into full-blown disputes[25].

The following issue resolution process is a typical approach[26]

- ❑ Resolve the problem at the lowest level of authority.
- ❑ Unresolved problems should be escalated upward by both parties in a timely manner, prior to causing project delays.
- ❑ No jumping of levels of authority is allowed.
- ❑ Ignoring the problem or 'no decision' is not acceptable.

> 'Partnering calls for providing as many opportunities for communication as possible. Traditionally a velvet curtain is often used by project team members to hide problems and hope they can resolve it themselves... Through an issue resolution process the stakeholders' experiences, both good and not so good, are put on the table. Risks and potentially difficult areas of the contract can be discussed openly. A high-trust culture is developed where everyone feels free to express ideas and make contributions to the solution[26].'

Partnering: overview

There may be many in the construction industries in the UK, Australia and New Zealand who view any management concept which emanates from the USA with a certain degree of scepticism, particularly if, as is the case with partnering, the concept is being promoted with a certain amount of evangelical hype. The promise of a win-win business environment may seem, to some, to be an unachievable utopia. However there are well-documented case studies[14,24,27] of many successes in the use of both project and strategic partnering. Bennett and Jayes[14] make the following comments:

> 'The initial investment in workshops and careful selection of partnering firms rapidly translates into significant net benefits that on individual projects can amount to 10% of total costs. Over time, strategic partnering can achieve the whole of Latham's 30% cost saving target.'

The CIIA[24] survey of partnering in Australia found that nearly 85% of respondents would undertake another partnering project

> 'Not surprisingly, partnering was considered a great success by the respondents who had experienced a partnered project with no contractual claims. Conversely, partnering was considered a failure when contractual claims amounted to more than 5%. In 56% of cases where partnering was a success there had been repeat business between partners. Where partnering had been unsuccessful there had been no repeat business.'

It is worth tempering enthusiasm for partnering with the findings of the CIIA survey which indicated that although partnering was seen as successful by the respondents and that those who had experienced successful partnering would partner again, on a five-point scale measurement of partnering success the mean of 3.83 was 'not overwhelmingly high'.

Given that partnering has to do with human relationships and trust, it can never be infallible. However the balance of evidence points in favour of partnering as a construction management concept which is capable of achieving a cultural shift from adversarial to non-adversarial, and hence it has the potential to significantly improve the net benefits for all the stakeholders. The principal benefits of partnering are generally seen as being[24]:

- Reduced exposure to litigation
- Improved project outcomes in terms of cost, time and quality
- Lower administrative and legal costs
- Increased opportunity for innovation and value engineering
- Increased chances of financial success.

A more detailed set of benefits was identified in the CIIA[24] study and was ranked in order of importance as follows:

(1) Exchange of specialist knowledge
(2) Reduced exposure to litigation
(3) Lower administration costs
(4) Financial success because of win-win attitudes
(5) Positive effect on claims costs
(6) Positive effect on schedule duration
(7) Better time control
(8) Better quality product
(9) Prompted technology transfer
(10) Better cost control
(11) Fostered innovation
(12) Reduced rework
(13) Evidence of innovation and improvement
(14) Improved safety performance
(15) More profitable job
(16) Fewer errors in documentation

(17) Innovation as a result of information exchange
(18) Mechanism for recording innovation.

In the introduction to this chapter we began with the comment that partnering is more than simply formalising old-fashioned values, or a nostalgic return to the good old days when a 'gentleman's word was his bond'. For partnering to be successful the ideals of partnering have to be implemented through the planning and implementation mechanisms which we have described.

> 'Critical performance indicators have a key role in this respect. Constant monitoring of performance is one of the most critical factors in achieving partnering success[24].'

In addition to the need to maintain good lines of communication, it is important that communications are diligently recorded so that all parties to the partnership are fully informed of the ongoing situation. Partnering however is not just about high ideals and creating an atmosphere of mutual trust; it also involves having good organisational systems in place and emphasising tasks of a routine nature, such as the keeping of proper records. For example respondents to the CIIA study rated the keeping of minutes at partnering meetings as the most important feature of continuous evaluation.

The concept of partnering is analogous to the torch of the Olympic movement. As the torch is passed from hand to hand, there is the ever-present danger that the torch will be dropped and the flame extinguished. Similarly, if the ideals of partnering are not continually revisited and carefully nurtured, the partnering message is likely to become distorted in the process of dissemination.

As stated at the beginning of this chapter, partnering is about human relations and human interactions. Partnering attempts to create a win–win situation for stakeholders by creating an environment of mutual trust. There is no guarantee however that all signatories to a partnering charter will be winners. The recent emergence of alliancing, which has been described as 'partnering underpinned with economic rationalism', is one attempt at solving this dilemma.

Project alliancing – a natural progression from project partnering?

Background

According to Doz and Hammel[28] one of the most dynamic features of modern corporate development has been the growth of alliances: 'Scarcely a day goes by without some significant new linkage being announced'. Alliancing is proving to be the prevailing sign of the flexible organisation and is in direct response to the changing corporate environment[29]. According to the Economist Intelligence Unit[30] alliances of all kinds will rank as one of the most significant management tools by the year 2010. Involvement in alliances is currently cited by 29% of EIU respondents, moreover this figure is predicted to escalate rapidly to no less than 63% by 2010[30]. Lewis[31], dubbed the 'corporate match maker', and acknowledged as one of the world's experts on alliances[32], predicts that the trend towards alliances will increase as more and more companies realise they cannot operate by themselves if they want to compete and be successful. However, as a cautionary note, it is worth bearing in mind that not all alliances are successful. The EIU[30] states that the success rate for alliances is about the same as marriage successes in the US, which is 50%.

Although the past decade has seen rapid and often discontinuous changes in market conditions in the global construction industry[33] there are, as yet, few examples of major project alliancing in the construction industry. This situation is likely to change with more and more construction companies attempting to achieve a balance between co-operation and competition in a highly competitive market place. Howarth *et al*[34] express the view that the inculcation of an attitudinal shift from adversarial to one of mutual trust and harmony can only be achieved through full co-operation and alliancing between the key participants in the industry. This comment rings true with many of the comments previously made in this chapter on partnering. How then does alliancing differ from partnering?

Alliance definitions

The scope and nature of alliances is reflected in the range of definitions which are in common currency. These definitions can be extremely broad such as 'A relationship between two entities, large or small, domestic or foreign, with shared goals and economic interests[35]' or 'organisations with capabilities and needs come together to do business and add value to the other partner, at the same time working to provide a product which enhances society and the capability of the ultimate client[36].' Other authors are more specific, for example: 'a cooperative arrangement between two or more organisations that forms part of their overall strategy, and contributes to achieving their major goals and objectives[33]' or 'a commercial collaboration between two or more unrelated parties whereby they pool, exchange or integrate certain of their respective resources for mutual gain while remaining independent'. Perhaps the clearest and most specific definition of the project alliance process is given by Gerybadze[37] who states: 'the client and associated firms will join forces for a specific project, but will remain legally independent organisations. Ownership and management of the cooperating firms will not be fully integrated although the risk of the project is shared by all participants.'

Despite this recent spate of definitions, alliancing is one of the oldest forms of business collaboration. The first alliance contracts were used in ancient Egypt when merchants co-operated in their commercial activities[38]. Bergquist[39] also notes that 'the pyramids of ancient Egypt were actually the result of quite sophisticated alliance arrangement'. The considerable increase in the number of alliances which have operated since the 1980s reflects the fact that corporations are becoming less reluctant to use this strategy, realising potential that had been demonstrated centuries before.

Alliancing in the construction industry

As previously discussed[33], although alliancing is experiencing exponential growth in the fields of commerce and business, examples of alliancing in the construction sector are few and far between. To date the closest connection that the construction industry has had with alliancing has occurred in energy and mining projects such as the Wandoo B oil pipe line and processing plant in Western Australia[40], Andrew Drilling, Platform Project, North Sea UK[41, 42] and the East Spar Development[41, 43]. The exception to the predominance of 'heavy' civil engineering type alliancing has been the recently completed Australian National Museum Project, in Canberra, which is claimed to be possibly a world first in terms of alliancing in a building context[41]. (The National Museum project has been the test bed for a series of research studies undertaken by the School of Construction Management and Property, Queensland University of Technology, the Department of Building and Construction Economics, Royal Melbourne Institute of Technology and CSIRO Building Construction and Engineering Division Melbourne. The papers emanating from this research[41, 44, 45] are commended to readers as an opportunity to study an emerging field of interest.)

The differences between alliancing and partnering

Walker *et al* discuss, at length, the differences between project alliancing and project partnering in their paper 'Project alliancing and project partnering – what's the difference? – Partner selection on the Australian museum project – a case study'[41]. In this paper they cite the following quotation from Thompson and Sanders[46] as exemplifying the differences between alliancing and partnering:

> 'While project partnering involves firms forging shared objectives and goals they still maintain a sense of independence with their own contractual arrangements and a tendering process that may or may not be based on a competitive cost structure. Partnering still involves a client buying a product – the project – through a procurement process that may involve any one of many forms (traditional, negotiated price, design and construction, management contracting etc.). Partnering, and its advantages, lies in attitudes and behaviours governing a commercial process. Project alliancing is different in that it is more all encompassing. It is better represented by coalescence.'

The key difference between partnering and alliancing is that in partnering all the signatories to the partnering charter do not necessarily gain from the partnering relationship. In some cases the partnering arrangement will be advantageous to some and disadvantageous to others. (It is worth noting in passing the C21 Conditions of Contract produced by the New South Wales Department of Public Works and Services. This set of conditions has been produced specifically for use on partnering projects and is designed to reduce the confrontational/adversarial culture which exists in conventional conditions of contract. This is an interesting example of a genuine response to the cultural demands of partnering. As previously discussed, the partnering charter has no legal standing. The C21 contract is an attempt to carry through partnering goals into a contractual relationship between the client and the contractor.)

The ethos of alliancing is somewhat akin to the slogan of the three musketeers 'All for one and one for all'. Alliancing could also be described as partnering underpinned with economic rationalism given that alliance partners coalesce into a *virtual* corporation[41] in which agreed profit and loss outcomes are contractually binding on all parties. ('A virtual corporation is an organisation that is separate from the contributing contractors and from the client, with an independent management structure and board[32]'). In a typical alliancing agreement, a body corporate will be appointed to provide management services to the project. The role of a body corporate is described by Gameson *et al* in their case study on the Wandoo B oil pipe line and processing plant in Western Australia[40]. The virtual corporation formed for this project was known as the Wandoo Alliance Proprietary Limited (WAPL). The role of WAPL was as follows:

- ❏ Provision of offices and general administration and support facilities for the Wandoo development
- ❏ Establishment and operation of bank accounts for the Wandoo development
- ❏ Procurement of services and materials as directed by the alliance board and/or the project director on instructions from the alliance board
- ❏ Other tasks designated by the alliance board and/or the project

director on instructions from the alliance board, or specified elsewhere in this agreement

The provision of funds, which are normally allocated via the client, were released to WAPL to enable WAPL to pay:

- Invoices submitted by parties
- Advances to alliance participants
- Invoices submitted pursuant to direct cost contracts entered into by WAPL
- Operating costs of WAPL in accordance with a budget approved by the Alliance board
- Payment of direct costs on behalf of Ampolex (the client)

Critical success factors in alliancing

There is a clear need to understand the appropriate conditions required for a successful alliance. Lynch[38] comments that 'Companies frequently lack sufficient management skills and resources to tackle extremely large and complex tasks.' The management team must especially have the required mindset capabilities for a successful collaboration[34]. Pivotal to the success of any project alliance is the selection of the appropriate partners[32]. Walker *et al*[41] explain the selection philosophy in the context of Australian National Museum as follows:

> 'In alliancing trustworthy, committed and world-class professional and competent firms are invited to join with the owner/client to develop the project. As an alliance of talented professionals pooling resources to achieve the project goal, they develop the project price target through design development with agreed risk and reward sharing arrangements established.'

In the case of the Wandoo B project[40] selection process and selection criteria were as follows. Ampolex (the client) contacted 19 companies by letter to respond to an expression of interest regarding the formation of an operator/contractor alliance to fully develop the

Wandoo field. The 19 companies contacted were chosen from technically and commercially prequalified contractors. The invited companies were asked to form their own prospective alliances in advance of discussions with Ampolex. Kava[47] notes that this decision was made because 'generally clients are not aware of past difficult working relations between various organisations. As the alliance is based on trust, not just between the client and the contractors but also between the contractors, any broken past relationship may hinder the trust'. It was conveyed from the very beginning of the alliance that Ampolex would not be a marriage broker.

Ampolex developed questionnaires and score sheets for use in selection interviews and meetings. The selection process was not rushed as it was thought to be important that the prospective alliance members understood the Ampolex approach to alliancing. Kava[47], the project integration manager of the Wandoo alliance, explains that the selection of the alliance team 'was purely a business assessment to answer the question: "could we meet our business objectives working in an environment based on trust?"'

At the core of Ampolex's selection process was the commitment by the CEO of each participant to contribute to an alliance as a virtual corporation to which skills, equipment and resources would be provided on an open book, net cost basis. Profit was separately negotiated. The successful group selected satisfied Ampolex's objective and subjective criteria which were as follows:

- ❑ All participants of a team were technically qualified
- ❑ A business assessment would be carried out on each company
- ❑ Each alliance group under consideration needed the capacity to jointly perform the whole of the scope of work
- ❑ The client should take no part in assembling the team
- ❑ Agreement was required on fundamental alliancing principles, including putting 100% of gross margin at risk
- ❑ There was an understanding that the project would proceed only if the owner's [Ampolex] financial criteria were met
- ❑ Reassurance was required of the total commitment of the CEO of each participant to alliancing concepts
- ❑ There should be no bidding process based on price (Wandoo B)

After the alliance was formed the negotiating of various agreements then commenced, and all members of the alliance team applied their collaborative flexibility, co-operation and trust. These provided the final reality check on the attitudes and commitments of the parties, and the documents were completed and executed in minimal time but with considerable contribution of effort by all participants of the alliance. The principles of the Wandoo alliance were:

- ❑ Trusting other team members and parties without waiting to see if they were trustworthy
- ❑ Communicating honestly and directly
- ❑ Listening generously to each other
- ❑ Supporting others and having mutual respect for others
- ❑ Accepting that it is OK to say no, but giving reasons; as opposed to saying yes, but not delivering
- ❑ Fostering an atmosphere of integrity, in which team members are expected to do what they say they are going to do
- ❑ Accepting and maintaining a level of stretch or discomfort in declared targets
- ❑ Being focused on achieving results
- ❑ Working on the basis of no blame is someone fails, especially if it is a failure to achieve a low probability stretch target
- ❑ Individuals taking ownership of their actions and inactions
- ❑ Working in an environment where problems are not seen as negative, but as avenues to new possibilities.

The financial arrangements

As previously emphasised, alliancing differs radically from partnering in respect to risk and reward sharing. In partnering the client still ultimately purchases a product (usually a building) which is produced, albeit in a spirit of mutual co-operation, with the design and construction team. In alliancing the virtual corporation produces the product with each member of the corporation sharing the risks and rewards, albeit that in most cases a percentage of any potential loss is underwritten or 'capped' by the client.

In the Wandoo B project the financial arrangements[40] contained two novel and critical elements:

- An open book approach
- Gainshare/painshare

Open book approach

One of the first financial arrangements that was agreed upon by the alliance participants was the 'open book approach' to the Wandoo project in which Ampolex provided details of Wandoo project economics to the alliance board.

Gainshare/painshare

The gainshare/painshare scheme placed as risk the aggregate of the profit and corporate overhead components of all parties, based on the overall performance of the alliance against the final cost. In the event of a cost overrun, the participants were to contribute 50% up to a maximum of the total of this profit and corporate overhead, but in the event of a cost saving, 50% of such saving was to be paid to the participants – with no cap, because there was no wish to diminish any incentive. No other penalty provisions were included because Ampolex understood that loss of gainshare, profit and overhead as well as the contribution to painshare were adequate incentives for performance.

The gainshare/painshare split among the parties was agreed based on a 50% allocation to Ampolex with the remainder divided in proportion to each party's contribution in the overall target cost. The gainshare/painshare split amongst the participants was agreed at alliance board level based on the percentage of cost that participants contributed to the project cost[47]. Figure 8.5 illustrates typical gainshare/painshare arrangements between alliance participants.

Note: if the Wandoo project was completed at less than target cost then additional profits were to flow to the participants in proportion to their gainshare/painshare percentage.

Note: if the Wandoo project overran the target cost then all the participants, including Ampolex, were liable for overrun in proportion to their gainshare/painshare percentage.

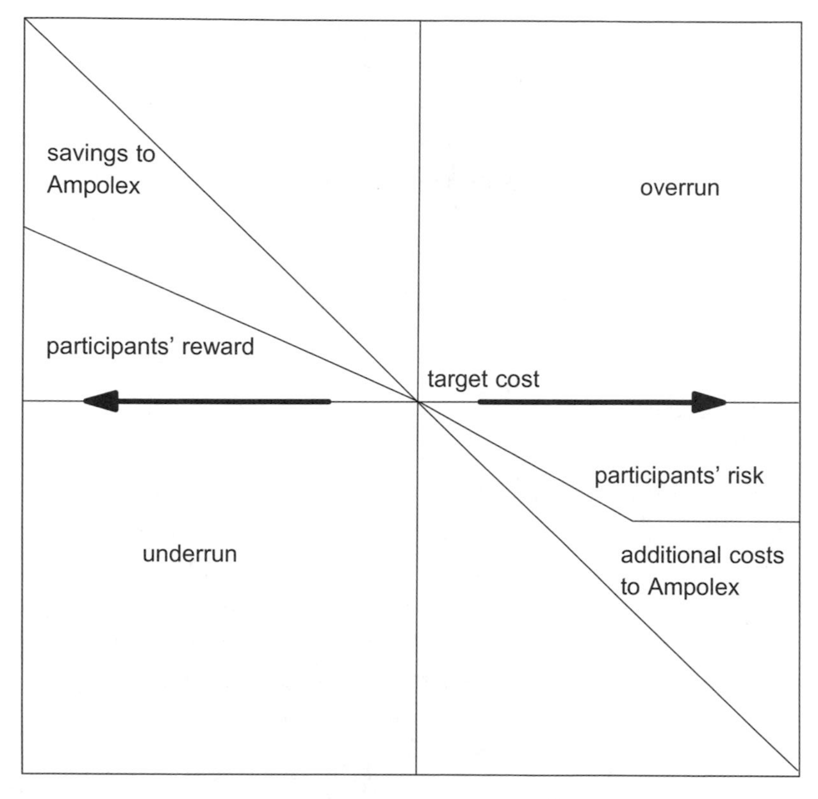

Fig. 8.5 Alliancing gainshare/painshare scheme[40].

The project outcome

Alliancing

The alliance has produced a platform and infrastructure that exceeded Ampolex's operational requirements. Oil production exceeded 8 million litres per day within three months of the first oil, 25% higher than the target production, and is continuing to exceed target levels.

Financially, the final direct cost of the development to Ampolex, excluding gainshare, drilling and production costs, has been limited to Aus. $364 million, some Aus. $13 million below the budget of Aus. $377 million established by the alliance board. The savings have been shared by the members of the alliance.

The Wandoo development was completed and achieved first oil

three weeks ahead of the Ampolex board's target date of 31 March 1997, and only 26.5 months after the project was sanctioned. The stretch target of 24 months was not achieved. Although the target was not achieved it is significant to compare this achievement to the findings of a recent worldwide benchmark study by Independent Project Analysis Inc. This study, covering 26 operators and some 300 projects, suggested that the oil industry norm for a similarly sized project was 34 months.

Postscript

In the Preface we stated that, in our view, alliancing is a natural progression from partnering. The recent success of the Australian National Museum project coupled with successes from the petrochemical field, such as the Wandoo B project, confirm our opinions. However it is significant that despite the increasing use of alliancing in the commercial sector, that alliancing has yet to be embraced as a management concept in the construction industry.

References

1. Hellard, R.B. (1995) *Project Partnering: Principle and Practice.* Thomas Telford, London.
2. Schutzel, H.J. & Unruh, V.P. (1996) *Successful Partnering: Fundamentals for Project Owners and Contractors.* John Wiley, New York.
3. Hinks, A.J., Allen, S. & Cooper, R.D. (1996) Adversaries or partners? Developing best practice for construction industry relationships. In: *The Organisation and Management of Construction: Shaping Theory and Practice.* (eds D.A. Langford & A. Retik) *Proceedings of the CIB W65 International Symposium*, 28 August–3 September. **1**, 220–28, Glasgow, Scotland.
4. Dubbs, D. (1993) Partnering means making friends not foes. *Facilities Design and Management*, June, 48.
5. Wright, G. (1993) Partnering pays off. *Building Design and Construction*, April, 36.

6. Kliment, S.A. (1991) Partnering makes sense. *Architectural Record*, March, 9.
7. Stasiowski, F.A. (1993) Partnering – the only approach for the 90s. *A/E Marketing Journal*, December, 1.
8. Hancher, D.E. (1989) *Partnering: Meeting the Challenges of the Future*. Interim report of the task force on partnering. Construction Industry Institute, University of Texas, Austin.
9. Godfrey, K.A. Jr (ed.) (1996) *Partnering in Design and Construction*. McGraw-Hill, New York.
10. Kubal, M.T. (1994) *Engineered Quality in Construction: Partnering and TQM*. McGraw-Hill, New York.
11. National Economic Development Office (1991) *Partnering: Contracting without Conflict*. June, NEDC, London.
12. Cowan, C., Gray, C. & Larson, E. (1992) Project partnering. *Project Management Journal*, December, 5–21.
13. Gyles, R.V. (1992) *Royal Commission into Productivity in the Building Industry in New South Wales*. Government of New South Wales, Sydney.
14. Bennett, J. & Jayes, S. (1995) *Trusting the Team: the Best Practice Guide to Partnering in Construction*. Reading Construction Forum, Reading.
15. New South Wales, Department of Public Works and Services. (1995) *Contractor Accreditation Scheme to Encourage Reform and Best Practice in the Construction Industry*. Government of New South Wales, Sydney.
16. Anonymous. (1995) Partnering and mediation gaining widespread acceptance, says survey. *Civil Engineering*, May, **65**(5), 28–30.
17. McManamy R. CII Benchmarks savings. ENR 1994 Aug.; 15.
18. New South Wales, Department of Public Works and Services (1995) *Capital Project Procurement Manual*. Government of New South Wales, Sydney.
19. Stevens, D. (1993) Partnering and value management. *The Building Economist*, September, 5–7.
20. Allot, K. (1991) Partnering – an end to conflict? *Process Engineering*, December, **72**, 27–8.
21. Katzenbach, J.R. & Smith, D.K. (1993) *The Wisdom of Teams: Creating the High-performance Organisation*. Harvard Business School Press, Boston, Mass.
22. Moore, C., Mosley, D. & Slagle, M. (1992) Partnering: guide-

lines for win-win project management. *Project Management Journal*, March, **23** (1), 18–21.

23. Kanter, R.M. (1989) Becoming pals: pooling, allying and linking across companies. *Academy of Management Executive*, **3**, 183–93. In: Cowan, C., Gray, C. & Larson, E. (1992) Project partnering. *Project Management Journal*, December, 5–21.
24. Construction Industry Institute Australia. (1996) *Partnering: Models for Success. Research Report 8.* Construction Industry Institute Australia.
25. Stephenson, R.J. (1996) *Project Partnering for the Design and Construction Industry*. John Wiley, New York.
26. Construction Industry Development Agency (CIDA). (1993) *Partnering: A Strategy for Excellence*. CIDA/Master Builders Australia.
27. Warne, T.R. (1994) *Partnering for Success*. ASCE Press, New York.
28. Doz, Y.L. & Hammel, G. (1998) *Alliance Advantage; The Art of Creating Value through Partnering*: Harvard Business School Press, Harvard.
29. de la Sierra, M. (1995) *Managing Global Alliances: Key Steps for Successful Collaboration*. Addison-Wesley, Reading, MA.
30. Economist Intelligence Unit (1997) *Vision 2010: Designing Tomorrow's Organisation*. EIU London/New York/Hong Kong.
31. Lewis, J. (1998) Developing trust in strategic alliances. *Channel Magazine* http://www.semi.org/Channel/1997/feb/features/trust.html.
32. Woods, B. (1997) The corporate match maker. *New Civil Engineer*, February, 16–18.
33. Kwok, A. & Hampson, K. (1996) *Building Strategic Alliances in Construction*. Queensland University of Technology, AIPM Special Publication.
34. Howarth, C.S., Gillin, M. & Bailey, J. (1995) *Strategic Alliances: Resource-sharing Strategies for Smart Companies*. Pitman Publishing, London.
35. United States Trade Center (1998) http://ustradecenter.com/alliance.html#introduction.
36. Nicholson, G. (1996) Choosing the right partner for your joint venture, Fletcher Constructions. *Proceedings of the Joint Venture & Strategic Alliance Conference*. 26–27 February, Sydney, Australia.

37. Gerybadze, A. (1995) *Strategic Alliances & Process Redesign*. Walter de Gruyter, New York.
38. Lynch, R.P. (1989) *The Practical Guide to Joint Ventures & Corporate Alliances*. John Wiley, & Sons Inc, New York.
39. Bergquist, W., Betwee, J. & Meuel, D. (1995) *Building Strategic Relationships*. Jossey-Bass, San Francisco.
40. Gameson, R., Chen, S.E., McGeorge, D. & Elliot, T. (2000) Principles, practice and performance of project alliancing – reflecting on the Wandoo B development project. *Proceedings of IRNOP IV Conference*, January, Sydney.
41. Walker, D.H.T., Hampson, K.D. & Peters, R.J. (2000) Project alliancing and partnering – what's the difference? – Partner selection on the Australian National Museum Project – a case study. In *CIB W92 Procurement System Symposium on Information and Communication in Construction Procurement*, Vol. 1 (ed. Serpell, A.), Pontifica Universidad Católica de Chile, Santiago, Chile, 641–55.
42. KPMG (1998) *Project Alliances in the Construction Industry*, NSW Department of Public Works and Services, Literature Review, 7855-PWS98-0809-R-Alliance, September, Sydney.
43. ACA (1999) *Relationship Contracting – Optimising Project Outcomes*, Australian Constructors Association, March, Sydney.
44. Hampson, K.D., Renaye, J.P. & Walker, D.H.T. (2001) Negotiation and conflict in innovative procurement environments: The National Museum of Australia. In *CIB World Congress*, Wellington, New Zealand.
45. Hampson, K.D., Renaye, J.P., Walker, D.H.T. & Tucker, S. (2001) Project alliancing and information technology in building construction: The National Museum of Australia. In *CIB World Congress*, Wellington, New Zealand.
46. Thompson, P.J. & Sanders, S.R. (1998) Partnering continuum. *Journal of Management in Engineering – American Society of Civil Engineers/Engineering Management Division*, **14** (5), 73–8.
47. Kava, J. (1996) No business as usual: an alliance action. *Partnering & Strategic Alliances Seminar*, the Institution of Engineers Australia Quality Panel, 22 November.

Chapter 9

Linking the concepts

Until this point in the text we have made no attempt to explore links between concepts (other than in Chapter 1 where we noted the holistic thread of systems theory). In the interests of clarity we have chosen to deal with each management concept as a discrete entity and have deliberately avoided the temptation to digress from one concept to the other. By this stage we hope that the reader will now have a good overview of all the concepts and have a clear understanding of the nature of each individual concept. Armed with this knowledge the reader is now equipped to explore the notion of the grey areas or fuzzy boundaries of adjoining concepts.

Our basic hypothesis which we stated in Chapter 1 is that:

> 'The probability is that the relatively slow and patchy uptake of modern construction management concepts is due not so much to a lack of diligence or a reluctance on the part of industry practitioners to adopt new ideas, but to the fact that these concepts need first to be understood and studied in total. Second, although government agencies are encouraging and, in some cases, attempting to enforce the adoption of the concepts, no guidance is being given on how the concepts can be applied concurrently and in combination. What is needed is a *weltanschauung* or world view based on a solid knowledge of the individual concepts.'

We are optimistic that this worldview is slowly emerging and that the adoption of the concepts covered in this text has and will be conducive to a cultural shift which will address the perennial problems of the construction industry of fragmentation[1,2,3], lack of integration and co-ordination between project participants[4] and the poor communication processes[5,6]. Whether this cultural shift will be achieved through the 'big bang' approach say of reengineering, the gentler approach of partnering or the all-embracing approach of total quality management is yet to be seen. In theory, at least, the concepts are not mutually exclusive: for example, it is possible to speculate on the scenario where a strategic partnering arrangement could in turn lead to a decision by the partners to reengineer all or parts of the procurement process, or conversely it is possible that partnering relationships could result from the implementation of reengineering. When we began to speculate on how linkages could be conceptualised we were drawn to the analogy of the Russian dolls where smaller and smaller versions of the same doll are nested one inside the other. It soon becomes clear, however, that the relationships between our concepts are much more complex than this and do not neatly fall into some sort of hierarchical order. For example we have already noted that partnering could be 'nested' inside reengineering or vice versa. If we then start to add successive overlays of concepts such as constructability and total quality management, then, while none of the concepts are necessarily mutually exclusive, a conceptual model of the inter-relationships becomes increasingly complex as successive concepts are added. Although there is certainly a commonality of purpose in all the concepts in terms of the breaking down of existing barriers, with the creation of different sets of relationships and lines of communication, the concepts all have differing goals and objectives which do not lend themselves to pigeon-holing. The solution to these problems lies, as we see it, in the proper application of the systems approach.

Many researchers in the field of construction management have advocated a systems approach. For example Kelly and Male[7] recommend the use of systems theory and systems thinking in the field of value management, as do Chen and McGeorge[8] in the development of a constructability model and we have also demonstrated how the systems approach underpins reengineering. Checkland[9] however wryly comments

> 'for some years now a systems approach has been a modish phrase. Few are prepared publicly to proclaim that they do not adopt it in their work and it would be an unwise author of a management

> science text who failed to sub-title his book: "a systems approach".'

There is much ambiguity about what the systems approach actually is. Often a systems approach is taken to simply imply a holistic view. Checkland observes however that

> 'the systems paradigm is concerned with wholes and their properties. It is holistic, but not in the usual vulgar sense of taking in the whole; systems concepts are concerned with wholes and their hierarchical arrangement rather than *the* whole.'

The interest for us lies in the notion of interrelationships of concepts.

The problem is how to apply systems thinking to construction management concepts in total rather than only to the specific concept domain. Part of the difficulty lies in finding a framework which accommodates the characteristics of the individual concepts within the system as a whole, which will avoid situations such as described by Sidwell and Francis[10] who, when identifying the unique characteristics of constructability, warn that 'constructability is not just value engineering or value management'. It is useful at this juncture to establish a basic description of systems thinking and the systems approach. Writers such as Armstrong[11] have described the systems approach as a useful and sensible way of organising thought. In Armstrong's major work on long-range forecasting he allocates a section of the work to 'The systems approach and other good advice'. He considers that there are two basic ideas behind the systems approach. 'First, one should examine objectives before considering ways of solving a problem; and, second, one should start by describing the system in general terms before proceeding to the specific.' The encouragement to first concentrate on the 'big picture' appears to be sound advice to both researchers and practitioners, and advice which would be difficult to refute. Armstrong's interpretation of how to operationalise the systems approach has been condensed in the form of a checklist as shown in Table 9.1.

At first glance, the checklist seems reasonable, and indeed fairly innocuous. The challenge, however, occurs immediately at step 1 (start at the highest conceptual level, use stakeholder analysis).

Table 9.1 Checklist for the systems approach (after Armstrong[11]).

Steps	Checklist of procedures
Identify objectives.	Start at the highest conceptual level. Use stakeholder analysis.
Develop indicators of success.	Build in reliability. Build in validity.
Generate alternative strategies.	Use experts from different areas. Use brainstorming.
Develop and select programmes.	Write scenarios.

Stakeholder analysis, as applied to an individual management concept, means identifying all the groups (or individuals) who are part of the system and who have an influence on, or are influenced by, the system. Stakeholder analysis also involves attempting to resolve the conflicting goals and objectives of the individual stakeholders. The problem, as we see it, is that many current construction management concepts appear to be in competition with one another for the attention of the same stakeholders. In other words, at a meta-concept level the first rule of systems theory is being broken, because no attempt is being made to resolve the relationship between the various concepts to the satisfaction of the stakeholders.

Barlow[12] developed a conceptual model which attempted to show relationships between management concepts as a series of islands surrounded by a sea. (The model was developed from an earlier conceptual model by Bjork[13] of IT islands and later modified by Stewart[14].) Figure 9.1 is a derivative of this model which can be used as a way of explaining the inter-relationships of current construction management concepts.

The conceptual model illustrates six management concepts as part of an archipelago appearing above the sea. As the surrounding sea recedes, more of each island appears and occasionally a completely new island emerges. On the surface, the islands do not appear to be linked; however, below the surface there is connectivity. By way of illustration TQM and partnering have been placed in close proximity to one another, as has value management and constructability with reengineering somewhat detached. The value of this conceptual model is that while acknowledging that the concepts belong to the

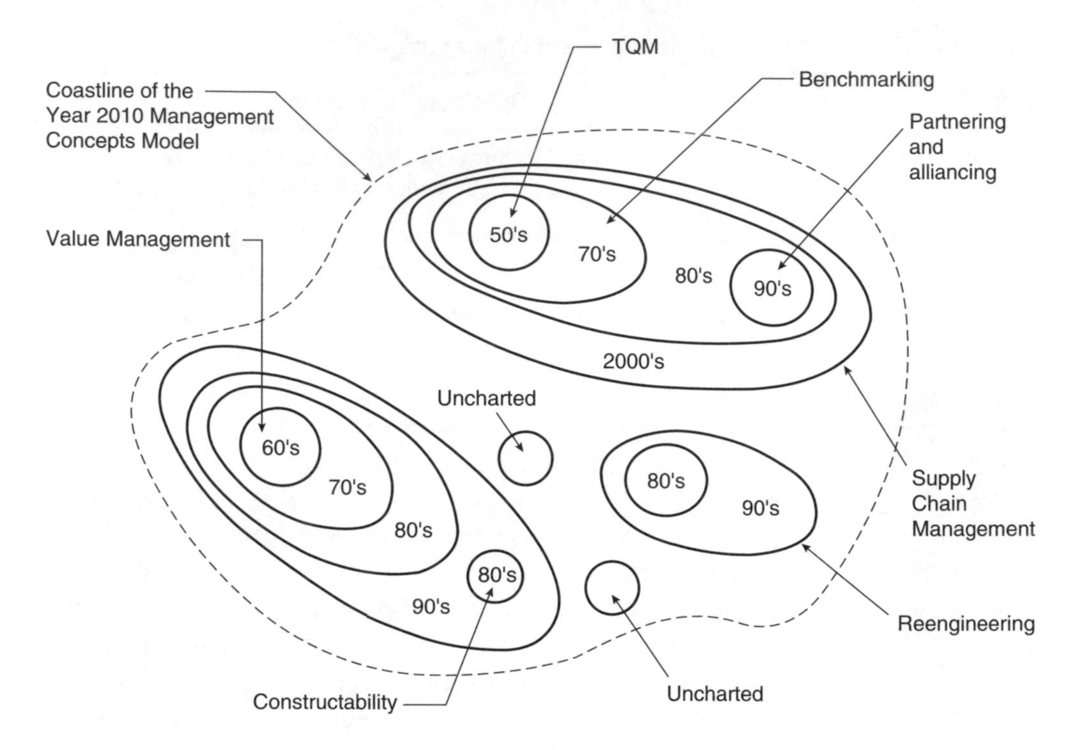

Fig. 9.1 Islands of management (after Bjork[13]).

same 'land mass', the model does not attempt to force or contrive a simplistic explanation of the linkage or impose a strict hierarchy. In the model there is an acknowledgement that much of what is below the water is as yet unknown. The model is also accommodating in terms of allowing for the emergence of new islands in, as yet, uncharted waters. To our knowledge, no group or organisation has yet attempted to simultaneously apply all these concepts. We predict that the industry trend, at least among the major players, will be in this direction. If and when this does occur, we can expect to see a hybridisation of current concepts which will minimise overlaps and fuzzy boundaries and will appear as a new archipelago on our island chart.

Since first proposing this model, we have received some criticism that the model gives the impression that emergence of management concepts is somehow a haphazard process subject to the vagaries of nature rather than part of a controlled development. We accept this criticism and on reflection agree that there is merit in this point of view. However we still hold to our archipelago notion based on the argument that serendipitous events do contribute to the emergence of new concepts and that it is much easier to identify patterns of development in hindsight than when attempting to predict the future. It could also be argued that supply chain management will introduce a wholly new dimension into relationships in the industry; however, this theory is yet to be tested.

The conceptual model is in line with the soft systems thinking using 'gentle guidelines' as advocated by Checkland[9]. Yeo[15] uses a similar analogy to the 'islands concept' when discussing the paradigm shift which occurred in the 1980s from hard to soft systems thinking, by stating that:

> 'The emphasis is now on the ability to interpret, and attach purpose and meaning to what is perceived in reality, and to actually understand what is actually happening *beneath the surface* in human affairs' (our italics).

He goes on to expand on this idea by stating that:

> 'The soft system methodology is not a technique, but a conceptualising process in which models representing human activity systems according to a particularly declared world view are constructed and used as a basis for comparison with the real-world actions in a problem situation. The model provides a framework to help generate insightful debates when comparing what should be done with what is actually happening in the real world.'

The concepts covered in this text are part of a process of a continuing cultural evolution (or revolution if one subscribes to the reengineering view). They are best viewed as being part of a fuzzy continuum, and in this respect the advice given by Yeo, that there is a need to look below the surface value, is well worth heeding. One of the pressing problems is the fact that there is very little case-study material available in the public domain in terms of feedback of user satisfaction of the current raft of construction management concepts. Godfrey[16] states that in the case of partnering, 'the use of partnering is growing fast, but there is a danger that it will be merely a passing fad. *User feedback, systematically recorded, is needed*' (our italics). This sentiment could equally be applied to all construction management concepts.

Useful work has been undertaken by the CII (Australia)[17] in collecting case-study material on the implementation of constructability on a range of projects throughout Australia. The constructability case-study data are particularly valuable because they can be analysed on the basis of the CII constructability systems model and constructability flow chart. The T40 case study[18] covered in Chapter 5 is a good example of the case-study approach which needs to be adopted on a wider front on live construction projects, as are the partnering case studies described in *Trusting the team*[19]. Ideally what needs to be undertaken are user-feedback studies of live projects which have a combination of several, if not all, of the concepts covered in this text. We would further advocate that the case-study data be structured on the basis of the 'islands' concept, or similar, which attempts to present a worldview of the relationships between the unique and the generic characteristics of in-vogue construction management concepts.

The questions which need exploring are:

- ❑ How can we establish mutually agreed indicators of success?
- ❑ Do practitioners see these in-vogue concepts as mutually compatible?
- ❑ Can a synergistic effect be observed when the concepts are used in combination?
- ❑ Alternatively, are some of the concepts mutually exclusive?

Clients are now responding to the urging of Latham and others and are starting to mandate the introduction of an ever-increasing number of construction management concepts through the use of prequalification criteria[20]. The fact that clients are now becoming more involved in decision-making in the construction process is to be applauded. However, unless practitioners can identify both the distinctions and the connectivity between management concepts, there is a real danger that the natural cynicism of the industry towards things seen either as modish or quasi-intellectual, will predominate and that what we will see is a token rather than a genuine commitment to cultural change.

References

1. Chen, S.E., McGeorge, W.D. and Ostwald, M.J. (1993) The role of information management in the development of a strategic model of buildability. *1st International Conference on the Management of Information Technology for Construction*. Singapore, August.
2. Leslie, H.G. and McKay, D.G. (1995) *Managing Information to Support Project Decision Making in the Building and Construction Industry*. CSIRO & NCRB, Melbourne.
3. RCBI (1992) *Productivity and performance of general building projects in New South Wales*. Report prepared by the Policy and Research Division, Royal Commission into Productivity in the Building Industry, Sydney, NSW.
4. Latham, M. (1994) *Constructing the Team*. HMSO, London.
5. Higgin, G. and Jessop, N. (1963) *Communications in the Building Industry: The Report of a Pilot Study*. Tavistock Publications, London.
6. Mortlock, B. (1980) Improved building communication. *Builder NSW*, 606–13.
7. Kelly, J. and Male, S. (1993) *Value Management in Design and Construction – the Economic Management of Projects*. E & FN Spon, London.
8. Chen, S.E. and McGeorge, D. (1993/4) A systems approach to managing buildability. *Australian Institute of Building Papers*, vol. 5.
9. Checkland, P. (1981) *Systems Thinking, Systems Practice*. John Wiley, Chichester.
10. Sidwell, A.C. and Francis, V.E. (1995) The application of constructability principles in the Australian Construction Industry. *Proceedings of the CIB W65 Conference*, Glasgow.
11. Armstrong, J.S. (1978) *Long Range Forecasting – from Crystal Ball to Computer*. Wiley-Interscience, New York.
12. Barlow, K. (1995) *Re-engineering – The next management revolution?* Unpublished thesis, Bachelor of Construction Management, The University of Newcastle, NSW.
13. Bjork, B. (1987) The integral use of computers in construction – the Finnish experience. *Proceedings of the ARECCAD Conference*, Barcelona, 19–21.

14. Stewart, P.J. (1995) *The impact of computer aided design and related technologies on the building procurement process*. Unpublished thesis, Master of Building, The University of Newcastle, NSW.
15. Yeo, K. (1991) Forging a new project value chain – paradigm shift. *Journal of Management in Engineering*, vol. 7, part 2, 203–12.
16. Godfrey, J.A. Jr (ed.) (1995) *Partnering in Design and Construction*. McGraw-Hill, New York.
17. Griffith, A. and Sidwell, A.C. (1995) *Constructability in Building and Engineering Projects*. Macmillan, London.
18. Ireland, V. (1994) *T40 Process Re-engineering in Construction*. Research Report, Fletcher Construction Australia Ltd, May.
19. Bennett, J. and Jayes, S. (1995) *Trusting the Team: the Best Practice Guide to Partnering in Construction*. Reading: Reading Construction Forum.
20. New South Wales, Department of Public Works and Services. (1995) *Contractor Accreditation Scheme to Encourage Reform and Best Practice in the Construction Industry*. Government of New South Wales, Sydney.

Bibliography

Value management

Miles, Larry. (1967) *Techniques of Value Analysis and Value Engineering*. McGraw-Hill, New York.

The first book to be written on value engineering and very much based on manufacturing. The text is useful in that despite the developments that have taken place Miles concentrates on value management as a concept as opposed to a technique. At times it appears that the technique has actually come full circle.

Dell'Isola, Al. (1988) *Value Engineering in the Construction Industry*. Smith Hinchman & Grylls, Washington DC.

A very practical text based primarily on the American 40-hour workshop approach. A lot of advice on how to do value engineering but thin on academic content.

Kelly, J.R. & Male, S.P. (1993) *Value Management in Design and Construction – the Economic Management of Projects*. E &FN Spon, London.

The only British text on value management in construction, this has a reasonable balance of academic and practical advice on value management. It includes methods of evaluation and a chapter on lifecycle costing.

The Europan Standard for Value Management EN 12973:2000.

This standard had a gestation period of nine years and is the result of collaboration between eight European countries. Although not definitive or restrictive, it is clearly a useful starting point in value management studies.

New South Wales Department of Public Works and Services (2001) *Value Management Guidelines*. Report no. 01054. Government of New South Wales, Sydney.

A detailed and comprehensive set of guidelines from the major Australian public works authority who have been practising value management for many years. The guidelines contain interesting case study material.

Constructability

Adams, S. (1989) *Practical Buildability*. Butterworths, London.

A UK perspective of constructability with case study material dealing with the relationship of design to construction.

Ferguson, I. (1989) *Buildability in Practice*. Mitchell, London.

A similar work to Adams also with a range of detailed case study material.

Griffith, A. & Sidwell, A.C. (1995) *Constructability in Building and Engineering Projects*. Macmillan, London.

A definitive work which comprehensively covers constructability in the UK, North America and Australia. It traces constructability from its origins in the Construction Industry Institute (CII) and explains the theory of the Construction Industry Institute Australia (CIIA) Principles File together with a series of case studies illustrating the application of these principles over a range of situations from strategic to detailed.

Benchmarking

Camp, R.C. (1989) *Benchmarking: The Search for Industry Best Practices that Lead to Superior Performance.* ASQC Quality Press, Milwaukee, Wisconsin.

The first and generally viewed as the definitive book on benchmarking by the inventor of the technique. It is geared towards manufacturing but its coverage of the topic is extensive.

Spendolini, M.J. (1992) *The Benchmarking Book.* American Management Association, New York.

Probably the 'best of the rest' of the books on benchmarking, it is more academic than Camp's work and is more comprehensive, having the benefit of being written later.

Copling, S. (1992) *Best Practice Benchmarking: The Management Guide to Successful Implementation.* Industrial Newsletters Limited.

This is typical of the other texts listed in the references in the benchmarking chapter. The works do not appear to add much to the subject but are merely the same information presented in different formats.

Peters, G. (1994) *Benchmarking Customer Services.* Financial Times/ Pitman Publishing, London.

This work is different in that it concentrates on only one aspect of benchmarking, namely customer focus.

McCabe, S. (2001) *Benchmarking in Construction.* Blackwell Science, Oxford.

A recent and welcomed addition to the body of literature on benchmarking. A particularly useful text as it is construction industry specific. The chapter on moving from theory to practice will be of particular interest to practitioners (although equally of value to students) in that it contains eight case study examples of benchmarking in the UK construction industry.

Construction Best Practice Programme (CBPP) website (2002) http://www.cbpp.org.uk/cbpp/index.jsp.

The website is frequently updated and, amongst other things, is a useful source of benchmarking and best practice activities in the UK.

Reengineering

Construction Industry Development Agency. (1994) *Two Steps Forward and One Step Back: Management Practices in the Australian Construction Industry*. CIDA, Commonwealth of Australia Publication, February.

Gives good contextual background on the construction culture in Australia.

Coulson-Thomas, C.J. (ed.) (1994) *Business Process Reengineering: Myth and Reality*. Kogan Page, London.

Drawing on a detailed, pan-European study of the experience, practice and implications of BPR, the book examines: the advantages and disadvantages of the technique as a tool for change; approaches and methodologies being used; success factors and implementation issues and the implications for organisations and those who work in them.

Hammer, M. & Champy, J. (1993) *Reengineering the Corporation: A Manifesto for Business Revolution*. Nicholas Brealey, London.

Co-authored by Michael Hammer, who is generally credited with initiating the reengineering movement. Practical advice, including case studies, on business process reengineering.

MacDonald, J. (1995) *Understanding Business Process Reengineering*. Hodder & Stoughton, London.

Explains the challenges and pitfalls surrounding building process engineering and provides a step-by-step guide to understanding the BPR process.

Morris, D. & Brandon, J. (1993) *Re-engineering your Business*. McGraw-Hill, New York.

Introduces the techniques of business process reengineering which bring about the changes in the very processes, practices, assumptions and structures upon which companies are built. It offers guidance on: how modelling and simulation techniques can be used to analyse current operations; the design of a new organisational structure that positions the company to take advantage of changes in the market-place; and the implementation of the new structure.

Petrozzo, D.P. & Stepper, J.C. (1994) *Successful Reengineering*. Van Nostrand Reinhold, New York.

Details how to implement a reengineering programme and what to avoid in the process. It stresses the evolutionary application of reengineering business processes, organisations and information systems. The text is divided into the four phases of a reengineering project in the order in which they will be carried out. The coverage within these phases includes: selecting the reengineering team; setting the project scope; understanding the current process and information architecture; preparing for redesign; redesign principles; and the final stage of reorganising, retraining and retooling.

CIB website http://www.cibworld.nl/pages/begin/Pro4.html.

The CIB covers a wide range of research and development interests. This site relates specifically to reengineering construction activities and is a useful source of ongoing CIB activities in this field.

Process Protocol Salford University website (2002)
http://www.Salford.ac.uk/gdcpp.

The website is regularly updated and is a useful source of information on the reengineering movement in the UK.

Total quality management

Vincent, K.O. & Ross, J. (1995) *Principles of Total Quality Management*. Kogan Page, London.

This text deals with the management of quality, process and quality tools, criteria for quality programmes, and case studies in quality. It also includes chapters on benchmarking and reengineering.

Evans, J.R. & Lindsay, W. (1993) *The Management and Control of Quality*. West Publishing Company, Minneapolis.

Covers similar material to the above but at the end of each section also gives review questions and practice problems.

Asher, J.M. (1992) *Implementing TQM in Small and Medium-sized Companies*. TQM Practitioner Series. Technical Communications Ltd, Letchworth, Hertfordshire.

Despite some information on managing quality in the medium-sized organisation the book contains fairly limited information on TQM.

Bounds, G., Yorks, L., Adams, M. & Ranney, G. (1994) *Total Quality Management: Towards the Emerging Paradigm*. McGraw-Hill, New York.

An extremely comprehensive text that deals with both the theory and practice of TQM.

Baden, H.R. (1993) *Total Quality in Construction Projects: Achieving Profitability with Customer Satisfaction*. Thomas Telford, London.

A text aimed purely at the construction industry. A significant portion of the text is devoted to auditing.

Choppin, J. (1991) *Quality through People. A Blue print for Proactive Quality Management*. IFS Publications, Bradford, UK.

This text deals with the people element of TQM but also has sections on the management of the process and of resources. It also presents a blueprint for the development of a quality programme.

Imai, M. (1986) *Kaizen. The Key to Japan's Competitive Success*. McGraw-Hill, New York.

A text devoted exclusively to *kaizen* and the Japanese art of management.

Supply chain management

AEGIS (1999) *Mapping the Building and Construction Product System in Australia.* Department of Industry, Science and Technology, Canberra.

An Australian federal government report outlining an attempt to view the construction industry from a different perspective. Many of the principles of supply chain management appear to underline the approach. Possibly one of the first to really attempt to understand the entire scope of the construction sector and all participants in the value chain, including clients/owners, property sector participants, maintenance firms, project related participants, site actors and offsite manufacturing and suppliers.

Ballard, G. & Howell, G. (2001) Lean Construction Institute website. http://www.leanconstruction.org/.

The Lean Construction Institute (LCI) was founded in August 1997 and is now a non-profit corporation. They conduct research to develop knowledge regarding project based production management in the design, engineering and construction of capital facilities. One of their principle aims is to extend to the construction industry the lean production revolution started in manufacturing. This approach maximises value delivered to the customer while minimising waste. Various construction companies contribute to LCI, support the research and participate in regular research meetings and implementation meetings.

Bowersox, D. & Closs, D. (1996) *Logistics Management: The Integrated Supply Chain Process.* McGraw Hill, New York.

A comprehensive overview of the supply chain management concept from the logistics perspective. A text suited particularly to higher level undergraduate or postgraduate students with useful exercises and problems.

Cox, A. & Townsend, M. (1998) *Strategic Procurement in Construction: Towards Better Practice in the Management of Construction Supply Chains,* Vol. 1, 1st edn. Thomas Telford, London.

The first comprehensive text relating strategic procurement principles to the construction supply chain. The chain is limited to the key client, contractor and consultant parties using six detailed case studies that represent best practice. An insightful critique on the factors that impact upon procurement approaches written largely from a political economy framework.

International Group for Lean Construction website
http://cic.vtt.fi/lean/.

A website reporting on the activities of the International Group for Lean Construction which includes both industry participants and researchers. Provides links to various other useful sites. Various key themes are addressed and a special page is devoted to supply chain research. It is an active site and one of the key activities of this group is the annual conference.

Nishiguchi, T. (1994) *Strategic Industrial Sourcing: The Japanese Advantage*. Vol. 1, 1st edn. Oxford University Press, New York.

A detailed account of the evolution of the Japanese subcontracting system in manufacturing industries. It traces the rise of subcontracting in three main periods: the rise of subcontracting through 1900–45, the post-war subcontracting period 1945–60 and the emergence of clustered control networks through the 1960–90 period.

O'Brien, W.J. (1998) *Capacity costing approaches for construction supply chain management*. Unpublished PhD thesis, Department of Civil and Environmental Engineering, Stanford University.

A leading edge doctoral thesis focusing on construction supply chain management.

Olsson, F. (2000) *Supply Chain Management in the Construction Industry: Opportunity or Utopia*. Licentiate in Engineering Department of Design Sciences, Logistics. Lund University.

A case study on the use of supply chain management in Sweden involving Skanska and IKEA.

Ross, D.F. (1998) *Competing through Supply Chain Management*. Vol. 1, 1st edn. Chapman & Hall, New York.

A text both useful to the critical practitioner and the academic, it explores many different aspects of fundamental principles and practical applications of supply chain management. It provides a methodology for implementation and also suggests areas for future research. It is aimed at mainstream management.

Riggs, D.A. & Robbins, S.L. (1998) *The Executive's Guide to Supply Chain Management Strategies: Building Supply Chain Thinking into all Business Processes*. Vol. 1, 1st edn. AMACOM, New York.

A text suited to industry practitioners, particularly CEOs or senior managers in service and manufacturing industries with useful strategic tools and techniques for supply chain management.

Partnering

Bennett, J. & Jayes, S. (1995) *Trusting the Team: the Best Practice Guide to Partnering in Construction*. Reading Construction Forum, Reading.

Particularly suited to industry practitioners about to embark on partnering or for those who wish to know more about the topic. Covers all aspects of partnering, making a clear distinction between project and strategic partnering, illustrated by the use of case study material.

Construction Industry Development Agency (CIDA). (1993) *Partnering: A Strategy for Excellence*. CIDA/Master Builders Australia.

Written by experienced partnering facilitators. Succinctly lays down the principles of partnering, includes pro formas from partnering charters, action plans, issue escalation process, performance objective evaluation system and rating forms for partnering exercises.

Construction Industry Institute Australia. (1996) *Partnering: Models for Success. Research Report 8*. Construction Industry Institute Australia.

A study of 16 partnered projects and two strategic alliances across Australia giving a balanced view of the benefits and disbenefits of partnering, with detailed coverage of the legal implications of partnering in an Australian context.

Cowan, C., Gray, C. & Larson, E. Project partnering. *Project Management Journal*, December, 5–21.

A journal article co-authored by Charles Cowan, one of the accepted gurus of the partnering movement, giving the basic principle of partnering.

Godfrey, J.A. Jr (ed.) (1996) *Partnering in Design and Construction.* McGraw-Hill, New York.

A recent North American Perspective on partnering which emphasises the relationships between contractor, subcontractor and owner.

Hellard, R.B. (1995) *Project Partnering: Principle and Practice.* Thomas Telford, London.

Lays particular stress on the need for cultural change and also covers partnering in Australia and New Zealand.

Schultzel, H.J. & Unruh, V.P. (1996) *Successful Partnering: Fundamentals for Project Owners and Contractors.* John Wiley, New York.

A recent North American perspective on partnering authored by two senior industry figures with origins in the Bechtel Company, a partnering pioneer.

Alliancing

Doz, Y.L. & Hammel, G. (1998) *Alliance Advantage; The Art of Creating Value through Partnering.* Harvard Business School Press, Harvard.

Prescribed reading as an introduction to alliancing. A much quoted text.

Kwok, A. & Hampson, K. (1996) *Building Strategic Alliances in Construction.* Queensland University of Technology, AIPM Special Publication.

Provides an introduction to alliancing in a construction industry context.

Gameson, R., Chen, S.E., McGeorge, D. & Elliot, T. (2000) Principles, practice and performance of project alliancing – reflecting on the Wandoo B development project. *Proceedings of IRNOP IV Conference*, January, Sydney.

A detailed description and a behind the scenes look at the first major civil engineering alliancing project in Australia.

Walker, D.H.T., Hampson, K.D. & Peters, R.J. (2000) Project alliancing and partnering – what's the difference? – Partner selection on the Australian National Museum Project – A case study. In *CIB W92 Procurement System Symposium on Information and Communication in Construction Procurement*, Vol. 1 (ed. Serpell, A.). Pontifica Universidad Católica de Chile, Santiago, Chile, 641–55.

The Australian National Museum project is claimed to be the first alliancing project undertaken on a building as opposed to a civil engineering project. This and other related papers (see text) is highly recommended reading for students and practitioners who wish to gain a building perspective on alliancing.

Index